Effective Techniques to Motivate Mathematics Instruction

Effective Techniques to Motivate Mathematics Instruction offers pre- and in-service teachers best practices and techniques that can be used to motivate students in the first few minutes of any lesson in mathematics. Veteran teacher educators Posamentier and Krulik show how a bit of creativity and planning up front pays back by enabling a successful lesson on even the most challenging mathematics topic. Organized around eleven different motivational techniques, each chapter includes a variety of illustrative examples of how the technique may be applied. Designed to complement any methods textbook, this updated and expanded version provides an accessible guide that helps future math teachers ease the transition from successful student to successful teacher by developing the tools needed to create motivational introductions in their classes. In-service teachers will also find ideas in this book that could rejuvenate their instruction.

Features and updates to the new edition include:

- ♦ Theory to practice focus helps ease future teachers' transition from mathematics methods courses to actual classroom scenarios. Each chapter offers illustrative examples of how motivational strategies can be applied to various secondary grade levels, from large classes to small group settings.
- ♦ A section on the use of technology for motivating mathematical relationships, focusing especially on the use of dynamic geometry computer programs and student engagement with technology to discover mathematical phenomena.
- ♦ A new section on the use of practical problems to motivate instruction. The authors suggest everyday examples and stories to promote student interest in a wide array of mathematical topics.

- A new Introduction addresses the challenges of motivating students in the age of standards-based mathematics, especially highlighting the impact of the National Council of Teachers of Mathematics' (NCTM) *Principles and Standards for School Mathematics* and *Common Core State Standards* (CCSS).

Alfred S. Posamentier is currently Chief Liaison for International Academic Affairs at Long Island University, New York. He is former Dean of the School of Education and Professor Emeritus of Mathematics Education at the City College of the City University of New York, and also held the same positions at Mercy College, New York. He was also Distinguished Lecturer at New York City College of Technology of the City University of New York.

Stephen Krulik is Professor Emeritus of Mathematics Education at Temple University, where he was responsible for the undergraduate and graduate preparation of mathematics teachers for grades K–12, and for the in-service training of mathematics teachers at the graduate level. Dr. Krulik received the Great Teacher Award from Temple University in 2002, and was given the Lifetime Achievement Award from the NCTM in 2011.

Effective Techniques to Motivate Mathematics Instruction

Second Edition

Alfred S. Posamentier and Stephen Krulik

Routledge
Taylor & Francis Group
NEW YORK AND LONDON

Second edition published 2016
by Routledge
711 Third Avenue, New York, NY 10017

and by Routledge
2 Park Square, Milton Park, Abingdon, Oxon, OX14 4RN

Routledge is an imprint of the Taylor & Francis Group, an informa business

© 2016 Taylor & Francis

The right of Alfred S. Posamentier and Stephen Krulik to be identified as authors of this work has been asserted by them in accordance with sections 77 and 78 of the Copyright, Designs and Patents Act 1988.

All rights reserved. No part of this book may be reprinted or reproduced or utilised in any form or by any electronic, mechanical, or other means, now known or hereafter invented, including photocopying and recording, or in any information storage or retrieval system, without permission in writing from the publishers.

Trademark notice: Product or corporate names may be trademarks or registered trademarks, and are used only for identification and explanation without intent to infringe.

First edition published by McGraw-Hill Education 2011

Library of Congress Cataloging-in-Publication Data

Names: Posamentier, Alfred S. | Krulik, Stephen.
Title: Effective techniques to motivate mathematics instruction / Alfred S. Posamentier and Stephen Krulik.
Description: Second edition. | New York, NY : Routledge, 2016. | First edition published by McGraw-Hill Education, 2011. | Includes index.
Identifiers: LCCN 2015046907 | ISBN 9781138640948 (hardback) | ISBN 9781138640955 (pbk.) | ISBN 9781315630854 (e-book)
Subjects: LCSH: Mathematics—Study and teaching. | Effective teaching. | Motivation in education.
Classification: LCC QA11.2 .P6384 2016 | DDC 510.71—dc23
LC record available at http://lccn.loc.gov/2015046907

ISBN: 978-1-138-64094-8 (hbk)
ISBN: 978-1-138-64095-5 (pbk)
ISBN: 978-1-315-63085-4 (ebk)

Typeset in Palatino
by Apex CoVantage, LLC

We dedicate this book to mathematics teachers who we hope will enlighten future generations so that they will learn to love mathematics for its power and beauty!

To my children and grandchildren, whose future is unbounded: Lisa, Daniel, David, Lauren, Max, Samuel, Jack, and Charles.

Alfred S. Posamentier

To my children and grandchildren, whose future is unbounded: Nancy, Daniel, Jeffrey, Amy, Amanda, Ian, Sarah, and Emily.

Stephen Krulik

Contents

Introduction xi

The Art of Motivating Students for Mathematics Instruction 1
What Is Motivation? 2

1 Indicate a Void in Students' Knowledge 11
Topic: The Introductory Lesson on the Tangent Ratio 12
Topic: Special Quadrilaterals 14
Topic: Determining the Measure of an Angle Formed
    by Two Secants to a Circle 15
Topic: Tangent Segments to the Same Circle 17
Topic: Introducing Heron's Formula for Finding the
    Area of a Triangle 18
Topic: Introducing the Quadratic Formula 19
Topic: The Introductory Lesson on Imaginary
    Numbers 20
Topic: Finding the Sum and Product of the Roots
    of a Quadratic Equation 22
Topic: The Introduction to Exponential Equations 23

2 Discover a Pattern 25
Topic: Counting Techniques 27
Topic: Introducing Non-Positive Integer Exponents 29
Topic: Caution with Patterns 30
Topic: The Sum of the Measure of the Interior Angles
    of a Polygon 33
Topic: Introduction to Counting Combinations 34

3 Present a Challenge 37
Topic: Introducing the Order of Operations 38
Topic: Determining Prime Numbers 41

Topic: Algebraic Applications 42
Topic: Introducing the Concept of π 44
Topic: Understanding the Value of π 45
Topic: Introducing the Circumference of a Circle 46
Topic: Finding the Sum of the Interior Angles
   of a Polygon 47
Topic: Proving Triangles Congruent 48
Topic: Introducing Geometric Series 51

**4 Entice the Class with a "Gee-whiz" Amazing
Mathematical Result** 53
Topic: Introducing the Nature of Proof 54
Topic: Thales' Theorem 56
Topic: Introducing the Nature (or Importance)
   of Proof 58
Topic: Considering Division by Zero 60
Topic: The Introductory Lesson on Sample Space
   in Preparation for Probability 61
Topic: Introduction to the Concept of Area,
   or Looking Beyond the Expected 63
Topic: Introduction to the Area of a Circle or
   to Finding Areas of Similar Figures 65
Topic: Infinite Geometric Series 66

5 Indicate the Usefulness of a Topic 69
Topic: Introduction to Proportions 70
Topic: Applying Algebra 71
Topic: Introduction to Similar Triangles 72
Topic: Introducing Modular Arithmetic 73
Topic: Introduction to the Concurrency of the
   Angle Bisectors of a Triangle 74
Topic: Determining the Volume of a Right Circular
   Cylinder 76
Topic: Introduction to Probability—Expected
   Outcomes 78
Topic: Introducing the Product of the Segments of
   Two Intersecting Chords of a Circle 79

Topic: Introduction to the Concurrency of the
Altitudes of a Triangle 80

6 **Use Recreational Mathematics** 83
Topic: Identifying Factors of Numbers 85
Topic: Understanding Percents 86
Topic: Reinforce Some Logical Thought in
Mathematical Work 87
Topic: Rationalize the Denominator of a Fraction 89
Topic: Applications of Algebra Explaining Arithmetic
Peculiarities 90
Topic: Applications of Algebraic Counterintuitive
Peculiarities 93
Topic: Introduction to Divisibility Rules, Especially
Divisibility by 11 95
Topic: Application of Algebraic Solutions to Digit
Problems 97

7 **Tell a Pertinent Story** 101
Topic: Introducing Divisibility Rules 103
Topic: Introduction to the Value of π 103
Topic: Introduction to Prime Numbers 105
Topic: Finding the Sum of an Arithmetic Series 107
Topic: Introduction to the Pythagorean Theorem 108
Topic: Introduction to the Centroid of a Triangle 110
Topic: Introducing the Law of Sines 112
Topic: Volume and Surface Area of a Sphere 114
Topic: Discovering a Prime Producing Function 116

8 **Get Students Actively Involved in Justifying
Mathematical Curiosities** 119
Topic: Introducing Probability 120
Topic: A Lesson on Digit Problems and Place Value 122
Topic: Application of Digit Problems in Algebra 123
Topic: Introducing the Base-2 Number System 126
Topic: Application of Digit Problems in Algebra, or
Using Algebra to Justify an Arithmetic Peculiarity 127

Topic: Applying the Trigonometric Angle
   Sum Function 128

9 Employ Teacher-Made or Commercially Prepared Materials 131
   Topic: Introducing the Concept of a Function 132
   Topic: Developing the Formula for the Area of a Circle . . . 133
   Topic: Developing the Sum of the Angles of a Triangle . . . 135
   Topic: Introducing the Triangle Inequality 137
   Topic: Extending the Pythagorean Theorem 138
   Topic: Introducing the Pythagorean Theorem 140
   Topic: Introduction to Angle Measurement with a
      Circle by Moving the Circle 142
   Topic: Concept of Similar Triangles 146
   Topic: Introducing Regular Polygons 147
   Topic: Introducing the Parabola 149

10 Using Technology to Motivate Mathematical Relationships 151
   Topic: Introducing Concurrency in Triangles 152
   Topic: Introducing the Properties of the Midline of a
      Triangle 152
   Topic: Introducing a Proof Requiring a Transformation . . . 154

11 Using Practical Problems to Motivate Instruction 157
   Topic: Introducing the Pythagorean Theorem 157
   Topic: Introduction to the Right Isosceles Triangle 158
   Topic: Introduction to Mean, Median, Mode 159
   Topic: Factors of Prime and Composite Numbers 160
   Topic: Introduction to the Area of a Non-right Triangle . . . 161
   Topic: Introduction to the so-called Hinge Theorem 162
   Topic: Introduction to the Odd-Even Properties
      of Numbers 163
   Topic: Introduction to the Sum of an Arithmetic
      Progression 164

Index 167

Introduction

In recent years, where the school curriculum has extended as a topic for discussion by teachers to now where parents are becoming more actively involved in their children's education, they, too, are concerned that the learning properly involves their own children. Towards that end the way a teacher should be motivating or engaging students in the topics of the curriculum has become a universal concern. This is particularly appropriate when many teachers are compelled to "teach to the test" for fear that their own position could be in jeopardy, since many states are now rating teachers by their students' performance on standardized tests. Hence, this book will provide the necessary ingredients as to how mathematics should be taught by motivating youngsters for each new topic to be learned.

The field of teaching mathematics always seems to be in a constant state of flux, constantly introducing new materials to be included in the curriculum. In the 1960s, we were faced with the so-called "New Math," which introduced us to concepts such as set language, symbolic logic, a more formal approach to geometric proofs and so on. In the 1980s, the National Council of Teachers of Mathematics (NCTM) gave us the *Agenda for Action*, which began by stating that "Problem solving must be the focus of the curriculum." By the year 2000 NCTM presented the *Principles and Standards for School Mathematics*. These began to suggest a different approach to algebra, for example, emphasizing patterns, relations and functions as well as using mathematical models to represent quantitative relationships. The Standards, as they were called, emphasized a strong algebraic foundation by the end of grade 7. This led to moving formal algebra down into grades 7 and 8, from its traditional place in grade 9. Many of the *concepts* of algebra were taught in the lower grades. The emphasis was on having our students master more mathematics before leaving high school. By 2010 we were faced with the *Common Core State Standards* (CCSS), familiarly known as the *Core Curriculum*. The emphases on the so-called common core curriculum were based on a series of standards. These standards were:

- Research and evidence based;
- Clear, understandable, and consistent;
- Aligned with college and career expectations;
- Based on rigorous content and the application of knowledge through higher-order thinking skills;
- Built upon the strengths and lessons of current state standards;
- Informed by other top-performing countries to prepare all students for success in our global economy and society.

These state standards were developed by committees empowered by governors and state commissioners of education. The core curriculum is supposed to be a "one size fits all" K-12 content curriculum for all schools, public and private throughout the United States and its territories. The emphasis is on learning "why" as well as how; on reasoning and problem-solving, as well as on skills. By 2013 the CCSS had been adopted by 45 states, although many of these states have been making changes during the last few years, and some are dropping these standards and reverting back to the previous standards. Many teachers feel uncomfortable with the curriculum generated by these standards, since their orientation and preparation was often quite inadequate.

Technology, too, has had an impact on *how* we taught as well as what was taught. Years ago, in the early 1960s, the overhead projector became a routine fixture in every classroom, permitting teachers to prepare complex drawings and proofs in advance, and to teach facing their students. As we moved on, the traditional chalkboard and chalk were quickly replaced by the whiteboard and markers. We were introduced to the calculator, which removed the emphasis on computation and refocused it on reasoning. To get an answer, all we had to do was push the appropriate buttons. But which buttons? That became the emphasis. The calculator was quickly joined by the computer. The technological advances never cease and constantly offer new methods to enhance instruction.

Yet, over the years and throughout all of this turmoil and change, one thing has remained constant. A good teacher always gets his or her students interested and involved in the lesson. They express a desire to participate. After all, if you can't get them interested, they won't learn! Motivation has always been the key element in planning any mathematics lesson. For years, teachers have sought ways to "grab" their students' interest, and make them want to become partners in the learning process, involved in learning what is being presented and taught.

We believe that the key to effective teaching is largely reliant on the teacher's ability to capture the genuine interest of the students for the material to be taught. This naturally rests on the planning that the teacher exerts in preparation for the lesson. Perhaps the single most important aspect of any lesson is the beginning of the lesson where the teacher must motivate the students for the ensuing lesson. This can be done in many ways and is also largely a function of the teacher's personality and voice. Studies have shown that what a teacher says accounts for 7 percent of the effectiveness package, the tone of the teacher's voice and the enthusiasm account for 38 percent, and the "body language" accounts for 55 percent. Teachers should be entertaining, without ever losing control of the lesson, and yet not be completely scripted to prevent accommodation to the quirks of any class.

Yet, even the finest style of presentation—an important part of any teaching performance—can only offer a portion of the overall effectiveness. The content of what is said is paramount! This then leads us into the theme of the book, namely, the techniques that can be used to motivate students in the first few minutes of almost any lesson in mathematics. This could be the most difficult part of a lesson to plan. It requires a modicum of creativity and yet it pays back by enabling a successful lesson. It is a very worthwhile investment of time.

From this brief introduction we expect the reader to quickly delve into the next section, The Art of Motivating Students for Mathematics Instruction—the real introduction to this very important aspect of teaching: motivating students and engaging them for the material to be learned.

Alfred S. Posamentier
Stephen Krulik
March 18, 2016

The Art of Motivating Students for Mathematics Instruction

Teaching an effective lesson should be the aim of every teacher every day. This presents a special challenge for the mathematics teacher, and an extra challenge for the secondary-school mathematics teacher, many of whose students are not terribly excited about the subject. Students need an exciting lesson—one that is carefully thought out and designed appropriately for each class. The beginning of a lesson, which not only sets the tone, but also can insure that the students will be receptive to the content to follow, is one of the most perplexing challenges—especially for new teachers: how to provoke an interest among students toward the ensuing class session.

For decades teachers have sought ways to do this effectively. Generally, without a definite formula for motivation, the best teachers can do is to begin the class in an interesting way and allow their genuine enthusiasm to be apparent throughout the lesson. This is often contagious and works to motivate the class. (Feigned or exaggerated enthusiasm, however, is easily perceived by students and could have a harmful effect on students' perceptions of the teacher.) What is needed is for teachers to develop—over time—an arsenal of motivational strategies to begin their mathematics classes. This can be done by simply building on the interests of the class and trying to relate these interests to the class session. It would appear that geometry, because of its visual nature, would readily generate interest among students if presented properly. Unfortunately, this is not always the case. Much of the high school course deals with proving theorems and then applying these theorems to what may be perceived as artificial problems. Those students interested in mathematics, in general, will probably be excited by this, as they will be interested by almost any mathematics activity. However, an effective teacher should focus much attention on the less interested students—the ones that need to become motivated to explore the topics being presented.

Rather than developing ideas for motivating students with topics of special interest to the teacher—often based on a teacher's personal experience—teachers should fortify themselves with a set of techniques from which motivational activities for almost any mathematics lesson can be drawn. This is precisely what this book is intended to provide: a collection of nine time-tested motivational strategies applicable to a variety of mathematical topics at almost all grade levels. The focus here will be at the secondary-school level, with a lot of illustrative examples provided for each motivational strategy. Remember, the objective here is to develop ways to make mathematics alluring.

What Is Motivation?

To motivate students is to channel their interests to the specific topic to be learned. This book will consider eleven techniques that can be used to motivate secondary-school students in mathematics. For each technique, examples will be presented that illustrate the wide variety of applications that can be used directly in the classroom, and, perhaps more importantly, illuminate the techniques for further teacher-developed applications. Guiding the teacher in developing other motivational devices based on the students' backgrounds and interests provides a skill that over time will prove to be a priceless instructional support.

Motivating students to learn mathematics is, for many teachers, the chief concern when preparing to teach a lesson. Students who become interested and receptive learners make the rest of the teaching process significantly easier and profoundly more effective. There are basically two types of motivation: extrinsic and intrinsic. *Extrinsic motivation* usually takes place outside the learner's control, in the learning environment, and, to a large extent, under the control of the teacher. *Intrinsic motivation* occurs within the learner and can be developed by the teacher, keeping several principles in mind. The techniques in this book were designed to provide meaningful and effective methods to intrinsically motivate students of mathematics.

When thinking of ways to generate student interest in a mathematics topic, certain *extrinsic* methods of motivation come immediately to mind, such as rewards that occur outside the learner's control. This may include token economic rewards for good performance, peer acceptance of good performance, avoidance of "punishment" by performing well, praise for good work, and so on. Extrinsic methods are effective for students in varying forms. Students' earlier rearing and environment have much to do with their adaptation of commonly accepted extrinsic motivators.

However, many students demonstrate intrinsic goals in their desire to understand a topic or concept (task-related), to outperform others (ego-related), or to impress others (social-related). The last goal straddles the fence between being an intrinsic and an extrinsic goal.

In a more structured form, *intrinsic* motivators tend to conform to the following basic types:

- *The Learner Wants to Develop Competencies.* Students are often much more eager to do a challenging problem than one that is routine. It is not uncommon to see students beginning their homework assignment with the optional "challenge for experts" problem, even if the time spent on this prevents them from completing the remaining items, which may be considered routine work.

- *The Learner Is Curious about Novel Events and Activities.* It is a natural human trait to seek out unusual situations or challenges that can be conquered by existing skills and knowledge and thereby provide a feeling of competence. When the learner's curiosity about unusual stimuli is piqued, it becomes a form of motivation.
- *The Learner Has a Need to Feel Autonomous.* The desire to act on something as a result of one's own volition is often a motivating factor in the general learning process. To determine for oneself what is to be learned, as opposed to feeling that learning is being done to satisfy someone else or to get some sort of extrinsic reward, is another basic human need.

The teacher's task is to understand the basic motives already present in the learners and to capitalize on these. The teacher can then manipulate this knowledge of students' motives to maximize the effectiveness of the teaching process. Often, this manipulation can result in some rather artificial situations, contrived specifically to exploit a learner's motives in order to generate a genuine interest in a topic. This is eminently fair and highly desirable!

With these basic concepts in mind, there are specific techniques, which ought to be expanded, embellished, and adapted to the teacher's personality, and, above all, made appropriate for the learner's level of ability and environment. The strategies should be taken to the classroom on a regular basis. They should be used as lesson starters and should be presented with an appropriate amount of teacher enthusiasm—a commodity that is essential with all aspects of teaching, but particularly appropriate when attempting to motivate the learner. This book will devote a chapter to each of the motivation techniques or strategies. After a thorough presentation of each strategy the chapter will provide ample illustrative examples from secondary grade mathematics levels, with the hope that these examples will enable the reader to create further applications of these strategies for the various classes being taught.

The first five to ten minutes of a mathematics class are usually perceived by many students as the time to comfortably prepare for the coming lesson. Unfortunately, too often, that time is used to make announcements, take attendance, and perhaps review the previous day's homework. For many students this is an open invitation to arrive late, socialize with their friends, or ignore any attempts at work. The "real" mathematics lesson starts once the class settles down. Furthermore, starting a lesson with these organizational activities often sets a poor tone for the rest of the class period. Teachers need some exciting and unusual activities to start a class to make the students want to come to class punctually. In addition to motivating students for the day's lesson, this initial period should set a positive tone—one of enthusiasm, and where possible, be used to demonstrate the power and beauty of mathematics.

Motivation Techniques

The following themes will be considered—one per chapter—as we explore the techniques for motivating mathematics instruction at the secondary-school level:

Indicate a Void in Students' Knowledge

Students usually have a natural desire to complete their knowledge of a topic. This motivational technique involves making students aware of a void in their knowledge and capitalizes on their desire to learn more. For instance, you may present a few simple exercises involving familiar situations followed by exercises involving unfamiliar situations on the same topic. Or you may mention (or demonstrate) to your class how the topic to be presented will complete their knowledge about a particular part of mathematics. The more dramatically you do this, the more effective the motivation. Guiding students to discover this void in their knowledge by themselves is most effective.

Discover a Pattern

Setting up a situation that leads students to "discover" a pattern can often be quite motivating, as students take pleasure in finding and then "owning" an idea. Some mathematicians find pattern discovery a key to their research. The trick is to use the pattern to develop the concept that is to be taught in that lesson.

Present a Challenge
When students are challenged intellectually, they react with enthusiasm. Great care must be taken in selecting the challenge. A problem (if that is the type of challenge used) must not only clearly lead into the lesson, but it must also be appropriate for the students' abilities. A challenge should be short and not complex. It should not be so engrossing that it detracts from the intended lesson. This would certainly defeat the purpose for which this challenge was intended. Thus, challenges may provide motivation for one class but not for another. Teacher judgment is important here.

Entice the Class with a "Gee-whiz" Amazing Mathematical Result
Unexpected results often intrigue the students and stimulate their curiosity. To motivate basic belief in probability, for example, discuss with the class the famous "Birthday Problem." Its amazing (and we dare say, unbelievable) result will have the class in awe and eager to pursue further study of probability, especially to justify these counterintuitive results.

Indicate the Usefulness of a Topic
Here a practical application is introduced at the beginning of a lesson. The application selected should be of genuine interest to the class, brief, and not too complicated, so that it motivates the lesson rather than detracting from it. Student interest must be considered carefully when selecting an application. Usefulness can be determined by the students' background knowledge of the topic involved in the application.

Use Recreational Mathematics
Recreational mathematics consists of puzzles, games, paradoxes, and other forms of mathematics entertainment. In addition to being selected for their specific motivational gain, these devices must be brief and simple. A student should achieve the "recreation" without much effort in order for this technique to effectively motivate a mathematical topic to be explored.

Tell a Pertinent Story
A story of a historical event or of a situation can motivate students. All too often, however, teachers, already knowing the story they are about to tell and eager to get into the "meat" of the lesson, rush

through the story. Such a hurried presentation minimizes the potential effectiveness the story may have as a motivational device. Thus, a carefully prepared method of presentation of a story for motivating a lesson is almost as important as the content of the story itself.

Get Students Actively Involved in Justifying Mathematical Curiosities

One of the more effective techniques for motivating students is to actively attempt to justify a pertinent mathematical curiosity. The students should be comfortably familiar with the mathematical curiosity before you "challenge" them to justify it. Although this could consume more time than may normally be allotted for a motivational activity, to proceed with a justification before sufficient exposure has been achieved would be counterproductive.

Employ Teacher-Made or Commercially Prepared Materials

Here, motivation can be achieved by presenting the class with concrete materials of an unusual nature. This may include teacher-made materials such as models of geometric shapes, geo strips, specifically prepared overhead transparencies, or practical "tools" that illustrate a specific geometric principle. Some fine commercially prepared materials are available, ranging from geometric models to videos of various kinds. Materials selected should be reviewed carefully and their presentation thoughtfully planned so as to motivate students for the lesson and not to detract attention from it.

Using Technology to Motivate Mathematical Relationships

The computer allows us to do things geometrically and otherwise that we were not able to demonstrate as effectively in earlier years. This device enables students to see and witness mathematical phenomena that then could leave them wondering "Is this always true?" Such a question can nicely lead into a desire for a justification or a proof—which might well be the topic of the lesson for which a teacher might have sought to generate interest—that is to motivate students!

Using Practical Problems to Motivate Instruction

Students often ask questions such as "What good is this?" or "Will I ever use this mathematics?" In this section we will suggest several real-life situations that make use of some mathematical relationships. The issue presented should require the mathematics that students think they already know to resolve the situation, but the illustration being presented here will demonstrate a more elegant way to resolve the issue. However, a word of caution. Teachers may not be able to use the specific models we propose in this book, since their students may not be familiar with the material or they simply may not be interested in the issues being considered. Rural environments breed other interests than urban environments. Teachers must be aware of the interests of their students and only use examples that will clearly pique their interests.

Going Forward in the Book
With each chapter focusing on one of these eleven motivational strategies, there will be an introduction to the technique presented, along with appropriate theory to solidify the ideas with the reader. This will be followed by illustrative examples of how the technique may be applied to the various secondary grade levels. For each example presented, there will be appropriate classroom application variations—large class and small group settings—as appropriate. In cases where there are mathematical concepts that may be a bit "off the beaten path" there will be ample justifications and explanations so that the reader will still feel comfortable using the technique.

As one becomes familiar with these eleven motivational strategies one must bear in mind some basic rules for implementing these techniques. We offer five such rules:

1. The motivational activity should be relatively brief.
2. The motivational activity should not be overemphasized. It should be a means to an end, not an end in itself.
3. The motivational activity should elicit the aim of the lesson from the student. (This is a fine way of determining how effective the motivation actually was.)
4. The motivational activity should be appropriate for the students' level of ability, readiness and interests.
5. The motivational activity should draw on the motives actually present in the students.

Secondary school mathematics teachers are always challenged to find ways to motivate their mathematics classes. There is unfortunately a societal displeasure with the study of mathematics, which sadly infects students at school. One of the best ways to combat this deleterious effect on the education of our youth is to motivate them towards this subject. What better way to do this than to make their mathematics instruction meaningful, which must begin with a motivated learner. That is the aim of this book. The textbooks today with all their fancy artwork and colorful designs often do not make the content match the appearance. Topics are usually presented in a straightforward way without concern for finding motivational techniques to introduce them. This is a task that is left for the teacher and one that often receives great praise from supervisors when they are done effectively. We cannot control a teacher's performance, but we can provide the teacher with the proper tools to develop on a regular basis one of the most important aspects of any lesson: the motivational introduction to a topic or a class session.

The content of school mathematics is always in flux. New topics are added, some are deleted, some are shifted around or placed in a different grade level. In the United States for example, the current trend is towards the "Common Core Standards." However, regardless of what new content the various curricula present, one of the most important aspects of successful teaching is to genuinely motivate students toward the various topics, concepts or material to be learned. Getting students interested and involved will always be of major importance.

1

Indicate a Void in Students' Knowledge

The desire to complete has always been part of human nature. A person is usually most satisfied when a task has been fully finished. Stamp collectors, for example, often spend years searching for that single elusive stamp that completes a particular set for their collection. Similarly, most students have a natural desire to complete a task or complete their knowledge of a particular topic. Awareness of these voids awakens a desire to fill them.

Beginning at a young age, children may inquire about a particular topic until it is fully explained. Although a simple answer may be sufficient at one time, more questions may arise at a later point, and a person may not rest until these, too, have been answered. This process is in reality lifelong, and possibly one of the most decisive drives that pushes humans to learn and develop. In the same way, students have a natural need to complete and expand their existing knowledge.

Often, it is very effective and motivating when a particular gap in knowledge is subtly indicated, such as with a topic that will be discussed to complete students' understanding of a specific area in mathematics. If a teacher acts cleverly, students can be made to discover this gap themselves, and thus their efforts to research this gap will increase until the gap is closed—their curiosity and attention have been awoken.

Naturally a teacher must be aware that it is important to reinforce recently covered subjects through well thought out examples before drawing attention to new topics, perhaps by indicating further knowledge gaps. It is also important to keep the curiosity and ambition of the students awake, so that formerly posed questions that may still have gaps can be discussed.

In this chapter we present some examples of how this motivational technique can be used for a variety of secondary-school topics. For instance, you might present a few simple examples involving familiar situations followed by an example involving an unfamiliar one, but on the same (or closely related) topic. Students should be led to the notion that they are actually lacking in a certain area where they previously (with the success on the first few familiar examples) felt comfortable. This tends to build a desire in the students to complete their knowledge of the topic at hand. This is different from telling students they are lacking information about a topic. Rather, through clever presentation, they are led to that feeling by themselves. This can be highly motivating, if done correctly. The more dramatically you do this, the more effective the motivation. Guiding students to discover this void in their knowledge plays a large role in making this motivational technique work well. We will examine a number of such motivators that lead the students to realize that they have reached a point where they must fill a void in their mathematical knowledge in order to complete their understanding of a topic.

Topic: The Introductory Lesson on the Tangent Ratio

Materials or Equipment Needed
Prepare the drawing below in advance, either on paper or computer, for presentation to the class at the start of the lesson.

Implementation of the Motivation Strategy
This activity is intended to motivate students for the first lesson on the tangent ratio. The design here is to indicate a void in the

students' knowledge about trigonometric functions without telling them that they are not familiar with this concept. Please bear in mind this is to be done without the aid of the Pythagorean Theorem.

Begin by having students find the value of x for each of the following:

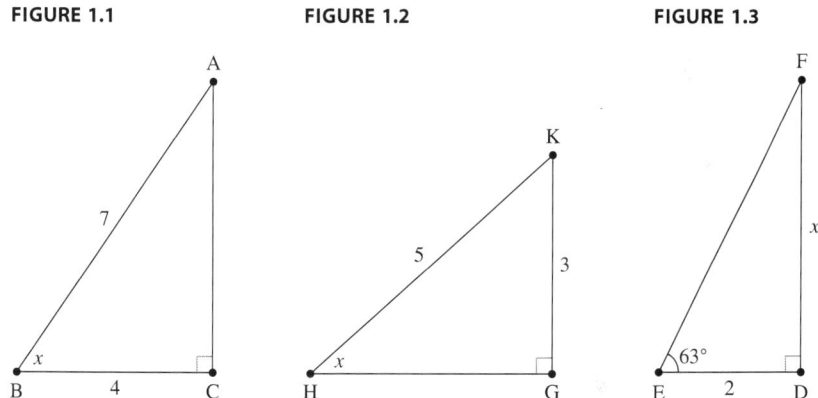

FIGURE 1.1   **FIGURE 1.2**   **FIGURE 1.3**

Students should have been introduced to the sine and cosine functions in an earlier lesson, and now they are to be motivated to learn about the tangent function. The value of x in figures 1.1 and 1.2 can easily be found by using the two trigonometric functions with which students should now be familiar. Namely, for figure 1.1, $\cos = x = \frac{4}{7}$, then $x \approx 55°$, and for figure 1.2, $\sin = x = \frac{3}{5}$, then $x \approx 37°$.

Presented properly, the students should now feel comfortable about finding an angle of a right triangle, where the sine and cosine function can be applied, that is, when the lengths of the hypotenuse and one leg are given. However, when confronted with the right triangle in figure 1.3, where the length of the hypotenuse is not given, while the length of one leg is given and the measure of an angle is also given, students will be made to realize that they cannot apply the cosine and sine functions to find the side of length x. (Please remember that the Pythagorean Theorem should not be used here.) Suddenly they sense there is a void in their knowledge of trigonometric functions. Presented properly, they ought to be motivated to fill this void.

They might realize that the teacher has this missing piece of information to fill this void—in this case, the tangent function. Applying the tangent function now allows students to determine the value of x in figure 1.3, namely: $\tan 63° = \frac{x}{2}$, then $x \approx 4$.

The technique of having students realize that there is a void in their knowledge base is effective, because of the natural tendency to complete an understanding of a topic. When it can be applied, it is particularly motivating and will have a favorable effect on the remainder of the lesson—in this case the study of the tangent function.

Topic: Special Quadrilaterals

Materials or Equipment Needed
A chart (figure 1.4), which should be on display for the class during their study of quadrilaterals.

Implementation of the Motivation Strategy
In the development of the various properties and descriptions of the special quadrilaterals, a chart such as that shown in figure 1.4 will give students a feeling of progressive or sequential

FIGURE 1.4

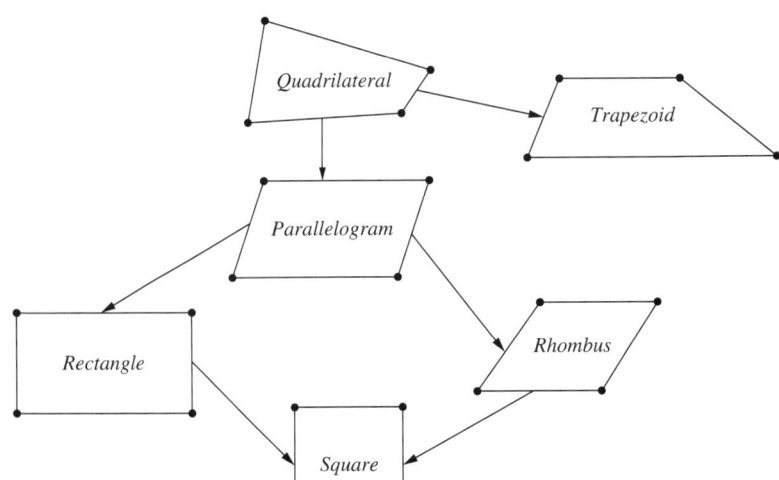

achievement, each time filling a void in their knowledge. This can be quite motivating as they anticipate each topic helping to come closer to completing their knowledge.

Students could be led to want to reach, sequentially, various levels of this diagramed development. The chart should be developed carefully, with the intended purpose clearly in focus.

Topic: Determining the Measure of an Angle Formed by Two Secants to a Circle

Materials or Equipment Needed
Chalkboard or other medium (e.g., Geometer's Sketchpad or GeoGebra) to present students with the following geometric problems.

Implementation of the Motivation Strategy
Students are often motivated by the sequencing of topics that clearly fit together and seem to support each other. For a class that has already learned the relationship of an angle formed by two chords intersecting inside or on the circle and its intercepted arcs, there could be an interest to determine the relationship of the measure of an angle outside the circle, such as one formed by two secants, a secant and a tangent, or two tangents. The motivation to complete the knowledge of this topic can be stimulated by the following questions: Find the measure of the angle marked x in the following three circles (see figure 1.5).

FIGURE 1.5

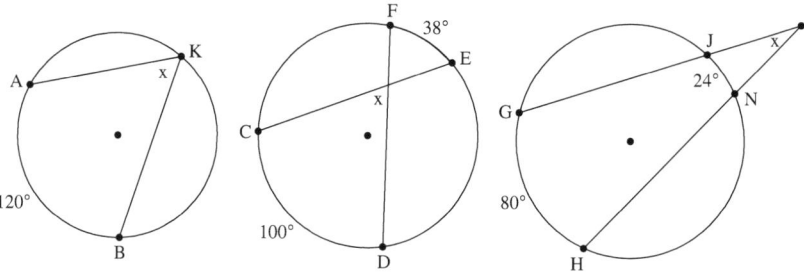

As you review the three questions after the class has had a chance to try to answer them, remind the class that they already know that an inscribed angle (figure 1.6) has half the measure of its intercepted arc.

Then remind them that they also know that the measure of an angle formed by two chords intersecting inside the circle (figure 1.7) is one-half the sum of the measures of the intercepted arcs.

However, they should be made to realize that to complete their knowledge of measuring angles related to the arcs of a circle they will have to include angles whose vertices are outside the circle. Figure 1.8 shows the types of external angles related to a circle.

With the feeling that they will now acquire the knowledge to complete the sequence, students will be motivated to determine how to find the measures of these angles from the measures of their intercepted arcs.

FIGURE 1.6

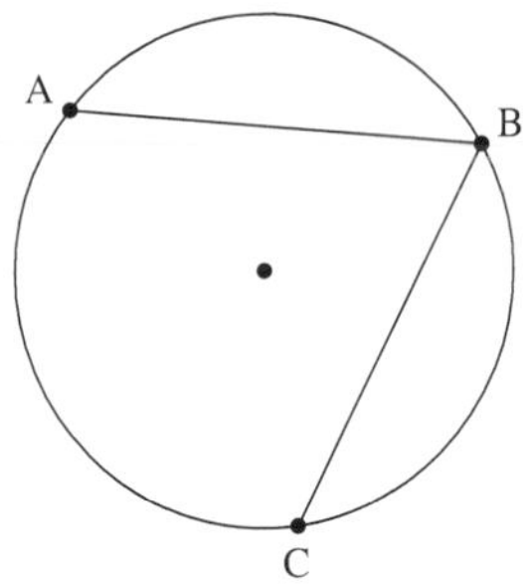

FIGURE 1.7

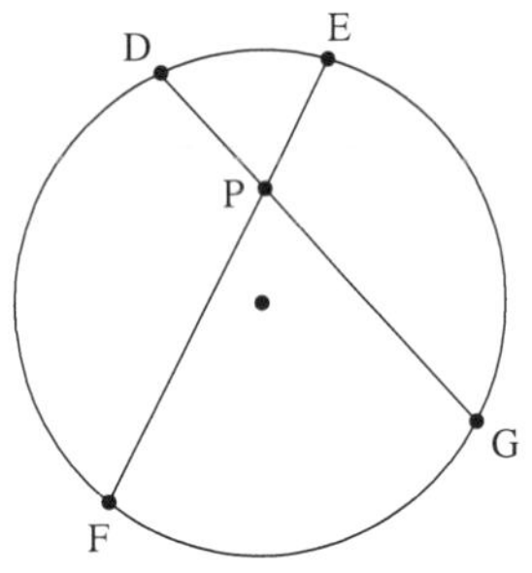

FIGURE 1.8

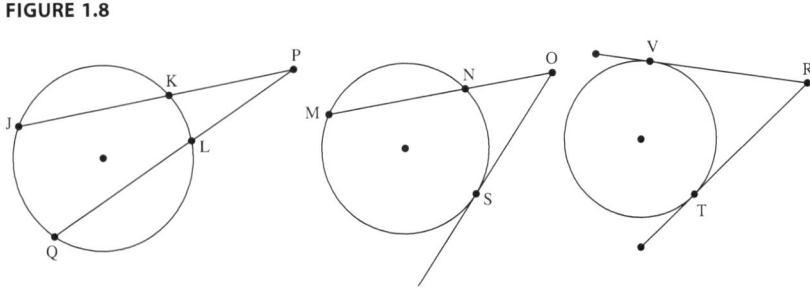

Topic: Tangent Segments to the Same Circle

Materials or Equipment Needed
Chalkboard or any other medium to present a problem to the class.

Implementation of the Motivation Strategy
Motivation can be created by a seemingly easy-to-understand problem that students may discover they are ill-equipped to solve, thereby causing them to realize they have a void in their knowledge. However, the solution of the problem hinges on a mathematical concept that will be explored in the ensuing lesson. The problem, presented in figure 1.9, where $\overline{PQ}$, $\overline{PR}$, and $\overline{TV}$ are all tangent to circle O at points Q, R, and S, respectively, asks students to find the perimeter of triangle PTV, if the length of tangent PQ is 8.

Intuitively, the students may feel that "something is missing," that it is impossible to solve the problem with only the given information. However, they need only know that tangent segments to a circle from a common outside point are equal, which is the lesson about to be taught.

The information to be learned follows: Since $\overline{TQ}$ and $\overline{TS}$ are both tangents from point T, they are equal in length, and we can replace TQ by TS. Similarly, $SV = VR$. Thus, the perimeter of triangle PTV consists of $PT + TS + SV + PV$.

FIGURE 1.9

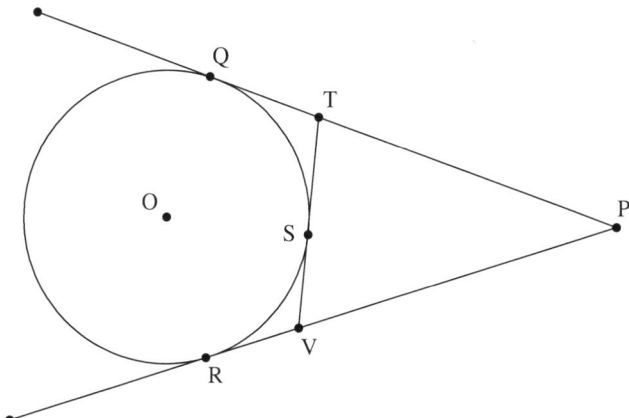

Since $PQ = PR$, the perimeter of the triangle is equal to the sum of the lengths of the two tangent segments from P, or $8 + 8 = 16$. The surprise of the simplicity of the solution, once the new theorem is presented, is a very motivating device.

Topic: Introducing Heron's Formula for Finding the Area of a Triangle

Materials or Equipment Needed
Chalkboard or any other medium to present students with the following geometric problems.

Implementation of the Motivation Strategy
The usual formula for finding the area of a triangle is $Area = \frac{1}{2}bh$, where b is the length of the base and h is the length of the altitude to that base. In geometry, there are other formulas for the area of triangles that students may already have learned. For a right triangle, the area is one-half the product of its legs, and for a triangle where the lengths of two sides and the measure of the included angle are given, the area is one-half the product of the lengths of the two given sides and the sine of the included angle (i.e., $Area = \frac{1}{2}ab \sin C$). However, how can we find the area of a triangle, if we are only given the lengths of its three sides? The answer is to use Heron's formula, which students can be motivated to discover by initially giving them the following problems to work out.

Present the following three problems at the start of the lesson:

1. Find the area of a triangle whose base length is 8 and height to that base is 6.
2. Find the area of a triangle whose side lengths are 3, 4, and 5.
3. Find the area of a triangle whose sides have lengths 13, 14, and 15.

For students to find the area of a triangle with base length 8 and height 6, they can easily use the formula $Area = \frac{1}{2}bh$, and find the area to be $\left(\frac{1}{2}\right)(8)(6) = 24$ square units.

Next, for students to find the area of a triangle whose sides are lengths 3, 4 and 5, they should recognize that this is a right triangle, with legs 3 and 4. Thus the area is given by $\left(\frac{1}{2}\right)(3)(4) = 6$ square units.

Now students are faced with the task of finding the area of a triangle whose sides are 13, 14, and 15. This is not a right triangle, so no side can be thought of as an altitude. There must be another way to find the area. Here the students should recognize a void in their knowledge of finding the area of a triangle.

One of the first student reactions is that, since 3-4-5 are sides of a right triangle, what about 13-14-15? Sadly, the triangle is *not* a right triangle. After trying to find the area (and failing to do so), the students should realize that a method must exist for finding the area of any triangle, given only the lengths of the three sides, and they hope that you can supply it for them.[1] This leads right into a lesson on Heron's formula for the area of a triangle. The formula was developed by the famous Greek mathematician, Heron of Alexandria[2] (AD 10–70).

The formula involves the use of the semi-perimeter, (s). The semi-perimeter is equal to $\left(\frac{1}{2}\right)(a+b+c)$ with a, b, and c being the lengths of the three sides. The formula is Area $= \sqrt{s(s-a)(s-b)(s-c)}$.

In this problem, $a = 13$, $b = 14$, $c = 15$. Therefore, the area of the triangle is $\sqrt{(21)(21-13)(21-14)(21-15)} = \sqrt{(21)(8)(7)(6)} = \sqrt{7056} = 84$ square units.

This will now give the students a feeling of filling the void that they may have felt at the start of the lesson—one that should have motivated them to learn Heron's formula.

Topic: Introducing the Quadratic Formula

Materials or Equipment Needed
Chalkboard, computer, or any other medium to present students with the following equations.

Implementation of the Motivation Strategy
The students have already learned how to use factoring to solve quadratic equations presented in various formats. This activity

will motivate them to recognize the need for a method for solving those quadratic equations when the quadratic polynomial is not factorable.

Give the students the following equations to solve. All of them can be solved by factoring except for the last one.

Solve for x:

(1) $x^2 + x - 6 = 0$ $[(x + 3)(x - 2) = 0; x = -3, 2]$
(2) $x^2 - 25 = 0$ $[(x + 5)(x - 5) = 0; x = +5, -5]$
(3) $2x^2 - 5x - 3 = 0$ $[(2x + 1)(x - 3) = 0; x = -1/2, +3]$
(4) $x^2 - 9x + 7 = 0$ [unfactorable over whole numbers]

The students should feel comfortable solving the first three equations, since each can be solved by a method with which they are familiar, namely factoring. Students should recognize that this last equation is not solvable by factoring over the whole numbers. This should lead them to ask if you are going to teach them how to solve this one in today's lesson. They have then perceived a void in their knowledge of solving quadratic equations, that is, those that are unfactorable. The quadratic formula, $x = \frac{-b \pm \sqrt{b^2 - 4ac}}{2a}$, will resolve their dilemma. Students are now receptive and hopefully eager to learn the formula.

You should point out to the class (after they have learned the quadratic formula) that this formula can be used to solve *all* quadratic equations, even those that can be solved by factoring. Go back and solve each of the first three equations on their worksheet using the formula. Show how the results are the same. They can then, finally, solve the fourth equation that frustrated them earlier in the lesson.

Topic: The Introductory Lesson on Imaginary Numbers

Materials or Equipment Needed
Chalkboard, computer, or any medium to present some equations for students to solve.

Implementation of the Motivation Strategy

Imaginary numbers—are they really imaginary? The entire concept of complex numbers and imaginary numbers is a difficult one for students to comprehend. Part of the problem is the nomenclature, i.e., "imaginary." The complex numbers have two parts: a "real" and an "imaginary." Yet the imaginary part is no more imaginary than the real part—it's just defining i as the $\sqrt{-1}$ to satisfy the rules of our number system. This motivator will indicate that there is a void in the students' knowledge when it comes to dealing with the square root of negative numbers.

Present the students with the following series of second-degree equations and ask them to solve for the variable.

Solve for x:

$x^2 - 1 = 0$
$x^2 - 16 = 0$
$x^2 = 0$
$x^2 - 17 = 0$
$x^2 + 9 = 0$
$x^2 + 27 = 0$

All goes well with the first four equations. The fifth and sixth equations in the set will pose a problem for them. Until now, any number squared must give a positive result. So, many will turn to their calculators to find the answer. Most calculators give a message of "ERROR" when the student tries to find the square root of −9 or −27. As students were given these equations to solve by the teacher, they must assume that there is some way to solve them, and that an answer can be found. This should lead to an introductory lesson on the use of "i," and the idea of complex numbers to fill the void in their knowledge that was just demonstrated to them through this set of equations.

As we said earlier, the textbooks usually refer to complex numbers as having two parts: a real and an imaginary part, written in the form $a + bi$ where a and b are real numbers and $i = \sqrt{-1}$. This is a completely arbitrary definition, and it is formulated so that the rules of our number system remain consistent. Thus, if we square i, we get $\sqrt{-1} \times \sqrt{-1} = i \times i = i^2 = -1$. This enables us to solve the last two equations:

$$x^2 + 9 = 0 \text{ yields } x = +3i \text{ and } x = -3i$$
$$x^2 + 27 = 0 \text{ yields } x = +5.196i \text{ and } x = -5.196i$$

Have your students substitute one of these values in each equation to show how they satisfy the original equation.

Topic: Finding the Sum and Product of the Roots of a Quadratic Equation

Materials or Equipment Needed
Chalkboard or any medium to present some equations for students to solve.

Implementation of the Motivation Strategy
Once your students have learned the formula for solving a quadratic equation, they should learn how to check to see if the roots they found are correct. While this can be done by substitution of the obtained roots into the original equation, the method of comparing the roots using the relationship of the sum and product of the roots to the original equation is far superior. This topic follows naturally from the quadratic formula; however we shall begin this motivational strategy with an indication of a need to learn the topic of the lesson.

Ask the students to solve each of the following equations, and then have them check the correctness of their results.

Solve for x and then check your results.

1. $x^2 - 4 = 0$ (Roots are +2 and −2.)
2. $x^2 - 10x + 21 = 0$ (Roots are 7 and 3.)
3. $x^2 - 7x + 5 = 0$ (Roots are 6.19 and .81.)
4. $x^2 + 3x - 16 = 0$ (Roots are 2.8 and −5.8.)
5. $x^2 - 4x - 7 = 0$ (Roots are 5.3 and −1.3.)
6. $2x^2 + x - 7 = 0$ (Roots are 1.6 and −2.1.)

The students should have no trouble with the first two equations. Both are factorable and yield integral answers that are easily checked. However, the third equation yields decimal answers, and checking via substitution back into the original equation is awkward and time-consuming. At this point, the students might expect to be taught another method, one which would be easier to implement, and the class is ready and perhaps motivated to do this checking another way. You have the "opening" to introduce the relationship of the roots of a quadratic equation to the coefficients of the original equation.

Once you have taught students the relationship between the coefficients of the equation and the sum and product of the roots, either give them the roots for the remaining quadratic equations or, if there is time, have them solve them and then check to see if the sum and product of these roots are as just developed: the sum $= -\frac{b}{a}$, and the product $= \frac{c}{a}$. Be sure to point out that the product may be off by a fraction of a decimal place due to rounding off the square root.

Topic: The Introduction to Exponential Equations

Materials or Equipment Needed
Chalkboard, computer, or any other medium to present a problem to the class.

Implementation of the Motivation Strategy
The following problem involving exponents may not look familiar to your algebra students but is probably well within their abilities. It is a good lead-in to exponential equations and a review of the laws of exponents and the simultaneous solution of a pair of equations in two variables.

Find the values of x and y, when $2^x = 8^{y+1}$, and $9^y = 3^{x-9}$.

While this appears to be a totally unfamiliar situation, students will feel that this is something that ought to be in their competence to solve. Hence, you have indicated a void in their knowledge. The solution is straightforward—once students realize that they need to express both sides of the equations with like bases as: $2^x = 2^{3(y+1)}$, and $3^{2y} = 3^{x-9}$. Since the bases are now equal in each equation, so are the exponents. This leads to the pair of equations:

$$x = 3(y + 1)$$
$$x - 9 = 2y$$

Solving these equations simultaneously yields $x = 21$ and $y = 6$.

Notes

1 Do not encourage students to draw the altitude to side length 14 and use the Pythagorean Theorem, to solve the problem, as it will ruin this motivational activity.
2 Some books refer to him as Hero of Alexandria.

2

Discover a Pattern

Mathematics is unique in its ability to give students a special kind of structure. Most people have a natural desire for structure, perhaps because it offers some sort of security, or something we can return to in the process of problem solving. As new knowledge is constantly added, there is a netting of crossover lines, patterns and drafts, which increases during our lives and makes it easier for us to handle new challenges as they arise.

In mathematics, new concepts can be made accessible from previous knowledge that has been at the students' disposal. If the teacher manages to cleverly set up a situation that leads students to discover a pattern, it can often be quite motivating. They feel great pleasure in finding and then owning an idea.

Naturally, the teacher can be an important supporter in this process, as he or she may carefully draw students' attention toward a somewhat hidden pattern. Done skillfully, the teacher's guidance ought to be so inconspicuous that students never lack the feeling of having discovered the idea for themselves. Such realization of their own ability is essential for the improvement of understanding and cognition.

Knowing different ways to resolve problems and having discovered patterns on our own provides tools, which can be used successfully in future contexts. A teacher should keep this in mind when considering merely writing a new formula on the board without allowing at least some modicum of student discovery. The reason for such an oversight might be a perceived lack of time and a focus on being able to promptly use this formula to deal with ensuing exercises. Often these formulas can be deduced in an easy and elegant way, using means that are at the students' disposal, notably by discovering a pattern.

Many of our students, and adults as well, have a distinct fear of mathematics. The only chance to fight against this widespread phobia is to awaken a deeper insight into the subject. But this can succeed only if instruction goes beyond imparting a body of rules and concepts, which can be recalled from memory as needed, to showing ways and structural approaches to mathematical tasks and exercises. This is nicely done through the recognition of patterns, which we'll demonstrate in this chapter with a variety of topics from the secondary school curriculum. This highly motivational technique is not always applicable, but where it is, it can be quite effective.

We encounter patterns every day. They pervade all aspects of our lives. Our ability to recognize patterns and to create them allows us to both discover and survive in the world. Many mathematicians consider patterns as the basis for all mathematics. Others consider patterns as a unifying theme. Revealing hidden patterns is what mathematicians do! The growth of mathematics has always been inspired by a search for patterns. Applications of mathematics often require the use of patterns to explain natural phenomena. Essentially, students go through several stages when discovering patterns. They first collect their data and try to organize it. They then attempt to classify the information they have collected. Once they have put their data into some kind of order, they try to find a pattern to explain the data. The more practice and experience the students have with searching for and finding patterns, the more adept they become at recognizing patterns from data. Problems that involve finding a pattern are numerous. However, you must search for those that might interest your students. Some patterns may not be mathematical in content, yet provide a great deal of motivation for the class. For example, you may ask your students to find the pattern rule for the following sequence:

$$8, 5, 9, 1, 7, 3.$$

Students may groan when they find the pattern rule, namely, the numbers are listed in order of the first letter of their spelling. They may feel similarly frustrated in discovering the pattern in the following sequence, when asked to find the next three terms.

$$1, 8, 3, 6, 5, 4, 7, 2, 9.$$

Here the sequence consists of two interspersed sequences of the ascending odd numbers and the descending even numbers.

Setting up a situation that leads students to "discover" a pattern can often be quite motivating, as we have said before. Students take pleasure in finding and then "owning" the idea. Students have been working with patterns since they were mere toddlers. A very young child separating marbles—"the big ones go here, the small ones go there"—is developing pattern recognition. As we mature, we observe and recognize more and more complex patterns. Observing a pattern and continuing to apply it can often lead directly to the resolution of a particular problem situation. Presenting activities and problems that lead to patterns is an excellent motivator for students in middle and senior high school grades. In this chapter, we will provide several motivators, all of which result in patterns or depend upon recognizing patterns to generate student interest.

Topic: Counting Techniques

Materials or Equipment Needed
The usual media to provide students with a problem (a computer projector would be best here, but a chalkboard would suffice).

Implementation of the Motivation Strategy

An important lesson for math classes at almost all levels is to demonstrate counting methods—those out of the ordinary. This session must be motivated by some dramatic examples. Here is one that uses finding a pattern. Begin by introducing a palindromic number—a number that reads the same forward and backward, such as 747 or 1991. Then ask students to determine how many palindromes are there between 1 and 1,000 inclusive.

The traditional approach to this problem would be to attempt to write out all the numbers between 1 and 1,000, and then see which ones are palindromes. However, this is a cumbersome and time-consuming task at best, and one could easily omit some of them.

Encourage students to see if they can find some helpful pattern. Counting the number of palindromic numbers in categories as shown in the table below should enlighten the class.

Range	Number of Palindromic Numbers	Total Number of Palindromic Numbers
1–9	9	9
10–99	9	18
100–199	10	28
200–299	10	38
300–399	10	48
...	...	...

Students should recognize a pattern: There are exactly 10 palindromes in each group of 100 numbers (after 99). Thus, there will be 9 sets of 10 = 90, plus the 18 from 1 to 99, for a total of 108 palindromes between 1 and 1,000. Pattern recognition is a highly motivating feature and should be used where possible to drive home important concepts and skills.

Another solution to this problem would involve organizing the data in a favorable way. Consider all the single-digit numbers (self-palindromes), which number 9. The two-digit palindromes also number 9. The three-digit palindromes have 9 possible "outside digits" and 10 possible "middle digits," so there are 90 of these. In total there are 108 palindromes between 1 and 1,000 inclusive.

Topic: Introducing Non-Positive Integer Exponents

Materials or Equipment Needed
Any of the usual media to provide students with a problem (a computer projector would be best here).

Implementation of the Motivation Strategy
Students who have just begun to understand the nature of positive integer exponents will probably respond to the question, "What does 5^n mean?" with a response such as "the product of n factors of 5." When asked what the nature of n is, they will probably say it is a positive integer. This motivator will encourage them to consider the non-positive integers: 0 and the negative integers.

You might show that these definitions enable an observed pattern to continue. Consider the following:

$$3^4 = 81$$
$$3^3 = 27$$
$$3^2 = 9$$
$$3^1 = 3$$

Then ask students to continue this pattern of dividing the result by 3 while decreasing the exponent by 1, to obtain:

$$3^0 = 1$$
$$3^{-1} = \frac{1}{3}$$
$$3^{-2} = \frac{1}{9}$$
$$3^{-3} = \frac{1}{27}$$

Students should use this pattern to motivate further examination of these negative exponents. They may also look at the issue by considering $\frac{x^5}{x^5}$ (where $x \neq 0$), which equals $x^{5-5} = x^0$. Therefore, $x^0 = 1$. However, the following is a meaningless statement: "x used as a factor 0 times is 1." To be consistent with the rules of exponents, we *define* $x^0 = 1$; then it has meaning.

In a similar way, a student cannot verbally explain what x^{-4} means. What would it mean to have "x used as a factor -4 times"? Using the rules of exponents, we can establish a meaning for negative exponents. Consider $\frac{x^5}{x^8} = \frac{1}{x^3}$. By our rules of operations with exponents, we find that $\frac{x^5}{x^8} = x^{5-8} = x^{-3}$. Therefore, it would be nice if $x^{-3} = \frac{1}{x^3}$, so we *define* it this way and our system remains consistent.

Once we arrive at $\frac{x^k}{x^k} = x^{k-k} = x^0$, we can now consider what value 0^0 should have. Using the same idea, we get $0^0 = 0^{k-k} = \frac{0^k}{0^k}$, which is meaningless, since division by $0^k = 0$ is undefined. Similarly, we cannot define 0^{-k}, for this yields $\frac{1}{0^k}$, which is undefined.

Thus, students can conclude that the base cannot be 0 when the exponent is 0 or negative.

In this way, the definitions $x^0 = 1$, and $x \neq 0$, $x^{-k} = \frac{1}{x^k}$, $x \neq 0$, become meaningful. The use of patterns motivates the students to keep exploring in this fashion to get a more solid understanding of exponents.

Topic: Caution with Patterns

Materials or Equipment Needed
The usual media to provide students with a problem—preferably a computer projector.

Implementation of the Motivation Strategy
The motivation for this topic is (naturally) a sequence, but one that is not expected. The topic of this lesson is recognizing when what appears to be an obvious pattern may not be the intended pattern.

A very common sequence such as this can be presented with the request to find the next term of the sequence: 1, 2, 4, 8, 16. When the next number is given as 31 (instead of the expected 32), cries of "wrong!" are usually heard. Yet: 1, 2, 4, 8, 16, 31 can be a legitimate sequence.

To show this we would need to further establish the sequence and then justify its existence in some mathematical way. It would be nice if it could be done geometrically, as that would lend a certain credence to a physical construct.

We set up a table (table 2.1) showing the differences between terms of the sequence), beginning with the given sequence up to 31, and then working backward once a pattern is established (here, at the third difference).

With the fourth differences forming a sequence of constants, we can reverse the process (turn the table upside down as shown in table 2.2), and extend the third differences a few more steps with 4 and 5.

The italicized numbers are those which were obtained by working backward from the third difference sequence. Thus the next numbers of the given sequence are 57 and 99. The general

TABLE 2.1

Original Sequence	1	2	4	8	16	31
First Difference		1	2	4	8	15
Second Difference			1	2	4	7
Third Difference				1	2	3
Fourth Difference					1	1

TABLE 2.2

Fourth Difference				1	1	*1*	*1*		
Third Difference			1	2	3	4	*5*		
Second Difference		1	2	4	7	11	*16*		
First Difference		1	2	4	8	15	26	*42*	
Original Sequence	1	2	4	8	16	31	*57*	*99*	

term is a fourth-power expression since we had to go to the fourth differences to get a constant. The general term (n) is:

$$\frac{n^4 - 6n^3 + 23n^2 - 18n + 24}{24}$$

Note that this sequence is not independent of other parts of mathematics. Let's examine the Pascal triangle (figure 2.1):

Consider the horizontal sums of the rows of the Pascal triangle to the right of the bold line drawn: 1, 2, 4, 8, 16, 31, 57, 99, 163. This is again our newly developed sequence.

A geometric interpretation can help convince students of the beauty and consistency inherent in mathematics. To do this we make a chart (table 2.3) of the number of regions into which a circle can be partitioned by joining points on the circle. Students should draw the circle and count the partitioned regions.

FIGURE 2.1

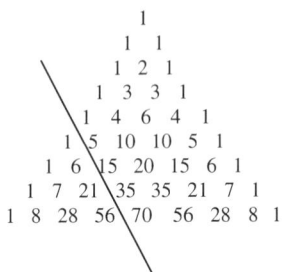

TABLE 2.3

Number of points on the circle	Number of regions into which the circle is partitioned
1	1
2	2
3	4
4	8
5	16
6	31
7	57
8	99

Now that students can see that this unusual sequence appears in various other fields, a degree of satisfaction may be setting in. However, the teacher should use caution when presenting students with a pattern so that it does not lead them astray.

Topic: The Sum of the Measure of the Interior Angles of a Polygon

Materials or Equipment Needed
A computer projector or other media to provide students with a problem; a geometry drawing software program such as Geometer's Sketchpad or GeoGebra would be useful.

Implementation of the Motivation Strategy
The motivational strategy of having students recognize a pattern can be very powerful and engender some deep understanding of the principle being introduced. Ask students to determine the sum of the measures of the interior angles of an icosagon (a twenty-sided polygon).

Instead of referring to a formula (which is itself derived from pattern recognition), or having students try various (often successful) attempts to answer the question, have them consider the angle sums of the polygons in increasing number of sides, listing their corresponding angle measure sums. Do they form a pattern? Is it easily recognizable? Can it be generalized? Can it be extended?

Have students begin with a triangle (interior-angle sum 180°) and then consider each of the polygons with successively increased number of sides, that is, quadrilateral, pentagon, hexagon, and so on. Students should be guided to find that they can triangulate the polygons by drawing lines from one vertex to each of the other vertices (see figure 2.2). When they do this, they will notice that each successive polygon includes one more triangle than its predecessor.

This yields a pattern which students should recognize and will lead to the desired goal. These angle sums should be put into tabular form so as to make it easier to recognize a pattern. This shows how a geometric pattern yields a numerical pattern.

FIGURE 2.2

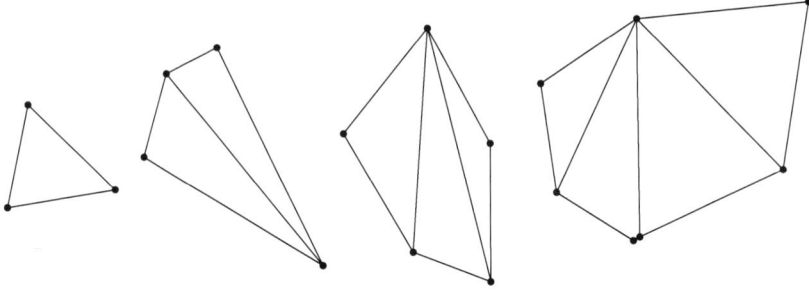

TABLE 2.4

Number of sides	3	4	5	6	7	8	9	...	20
Number of triangles	1	2	3	4	5	6	7	...	18
Angle measure sum	180	360	540	720	900	1080	1260	...	3240

An inspection of the first seven entries in table 2.4 (although we really didn't need that many) shows a pattern; namely, when the number of sides increases by 1, the number of triangles increases by 1 and the angle sum increases by 180°. Thus, for a nonagon (9-sided polygon) the number of triangles formed would be 7, and the angle sum would be $7 \cdot 180° = 1260°$. Using this pattern, we could work our way up to the 20-sided polygon we seek, using the pattern of increments of 180°, or we could recognize the pattern as one which implies that, for the 20-sided polygon, the angle measure sum will be 180° times 18 (which is 2 less than the number of sides). Thus for the icosagon, the angle measure sum is $18 \cdot 180° = 3240°$.

This motivational activity involving exhibiting a pattern served the dual purpose of motivating students while developing the main theme of the lesson.

Topic: Introduction to Counting Combinations

Materials or Equipment Needed
A computer projector or other media to provide students with a problem; a geometry drawing software program such as Geometer's Sketchpad or GeoGebra would be useful.

FIGURE 2.3

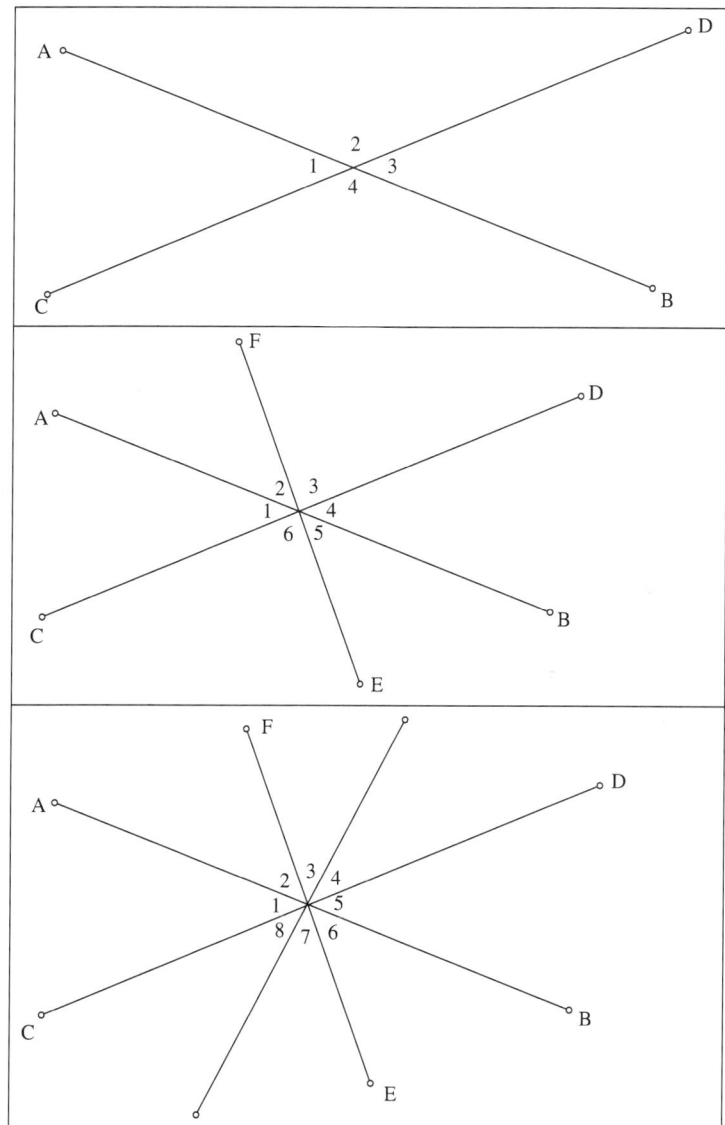

Implementation of the Motivation Strategy

Begin with the following problem for the students to ponder and then answer:

> How many pairs of vertical angles are formed by 10 concurrent distinct lines?

Students will often attempt to draw a large, accurate figure, showing the 10 concurrent lines. They will then attempt to actually count the pairs of vertical angles. However, this is rather confusing, and they can easily lose track of the pairs of angles under examination. You will now have the chance to motivate them with the demonstration of a pattern that evolves from this challenge.

Have the students consider starting with a simpler case, perhaps only two lines intersecting, and then gradually expanding the number of lines to see if a pattern emerges (see figure 2.3). If we start with 1 line, we get 0 pairs of vertical angles.

Two lines produces 2 pairs of angles: marked as 1–3 and 2–4 in figure 2.3.

Three lines produce 6 pairs of angles: they are 1–4; 2–5; 3–6; (1,2)–(4,5); (2,3)–(5,6); (1,6)–(3,4).

Four lines produces 12 pairs of vertical angles: 1–5; 2–6; 3–7; 4–8; (1,2)–(5,6); (2,3)–(6,7); (3,4)–(7,8); (4,5)–(8,1); (1,2,3)–(5,6,7); (2,3,4)–(6,7,8); (3,4,5)–(7,8,1); (4,5,6)–(8,1,2).

Students now ought to be able to summarize the pattern they have discovered, as in the table below:

Number of Lines	1	2	3	4	5	...	n
Pairs of Vertical Angles	0	$2 = 2 \cdot 1$	$6 = 3 \cdot 2$	$12 = 4 \cdot 3$	$20 = 5 \cdot 4$	...	$n(n-1)$

By now students should realize that for 10 distinct lines, there will be $10(9) = 90$ pairs of vertical angles.

This should lead into the lesson on how combinations can be counted more automatically. Students can also consider this problem from another point of view: Each pair of lines produces 2 pairs of vertical angles. Thus we ask how many selections of 2 lines can be made from 10. The answer is, of course, $_{10}C_2 = 45$. Thus we get $45 \cdot 2$ or 90 pairs of vertical angles.

3

Present a Challenge

Who cannot remember being enticed by an intellectual challenge so consuming that one becomes obsessed with finding a solution? Solving a problem through perseverance and cleverness may be one of the most rewarding intellectual experiences. To offer students a chance to solve such a problem requires the teacher to select an appropriate challenge. The better that teachers know his or her students, the easier it will be to select the appropriate problem with which to challenge the class. Naturally, the problem should not be so easy that it misses the challenging effect, nor so difficult that it leaves students frustrated. Above all, the challenge should lead to the lesson for which this is the intended motivation, and should result in the students being ready, curious, and open for the ensuing lesson. It should not detract from the lesson.

Teachers who have heterogeneously grouped classes might find it advantageous to create groups of two or three students to consider the challenge presented to the class; larger groups often prove counterproductive as they invite some students to sit passively while others are actively engaged in attacking the challenge. Furthermore, having several small groups might result in alternative solutions to the presented problem, providing different and unexpected entrances to the topic of the intended lesson. Additionally, within the small groups, individual student challenges can produce some creative results. These can differ significantly from an adult's reasoning process and yet be quite useful as an entrance point to the lesson.

It is not always easy to select the proper challenge which will serve as a motivational device for a particular lesson. Yet there are sources other than textbooks which provide interesting, challenging problems, which, if used properly, can be presented succinctly so as to require relatively little time for students to comprehend. Thus, they would provide appropriate motivation for students to lead into the topic of the ensuing lesson.

Topic: Introducing the Order of Operations

Materials or Equipment Needed

A worksheet, as shown in table 3.1, or these items projected or written on the board.

Implementation of the Motivation Strategy

Students in any grade in middle school where the order of operations is being taught can benefit from this activity. The motivational device is to present a challenge for students to show that they can acknowledge the order of operations. The activity demonstrates to the class that the order of operations can yield different answers if incorrectly applied. Some of the operations students can use may not be taught until high school; thus the activity can be revisited at any grade level.

TABLE 3.1 Worksheet

1.	4 4 4 4	Target: 2
2.	4 4 4 4	Target: 36
3.	4 4 4 4 4	Target: 16
4.	4 4 4 4 4	Target: 0
5.	4 4 4 4 4	Target: 7

Begin the motivational activity by providing each student or group of students with the sheet shown in table 3.1. The sheet shows some groups of 4s, and the students are asked to obtain each of the natural numbers shown as a target using only the operation signs +, −, ×, ÷ and exponents, as well as parentheses where appropriate.

As students work on the problems, they will notice that some students in a group may arrive at a "target number" in different ways. Some may not even be able to arrive at the target number at all. Students must decide what grouping symbols and what operational symbols can be used to arrive at the given target number. There may be different solutions which yield the correct target number. Be sure to discuss each solution (correct or incorrect) carefully. Indicate where the incorrect order of operations was used. This activity creates the need for a lesson on the definitive order of operations, in effect, motivating students to get the "rules" right.

Here is one set of possible answers. Note that others may also satisfy.[1]

1. $(4 \div 4) + (4 \div 4) = 2$
2. $4 \times (4 + 4) + 4 = 36$
3. $4(4 + 4) - (4 \times 4) = 16$
4. $(4 - 4) - 4(4 - 4) = 0$
5. $\sqrt{4} + \sqrt{4} + \sqrt{4} + (4 \div 4) = 7$

For the ambitious teacher wanting to provide the class with some enrichment, we offer how four 4s can be used to express the natural numbers from one to twenty.

$$1 = \frac{4+4}{4+4} = \frac{\sqrt{44}}{\sqrt{44}}$$

$$2 = \frac{4-4}{4} + \sqrt{4}$$

$$3 = \frac{4+4+4}{4} = \sqrt{4} + \sqrt{4} - \frac{4}{4}$$

$$4 = \frac{4-4}{4} + 4 = \frac{\sqrt{4 \cdot 4} \cdot 4}{4}$$

$$5 = \frac{4 \cdot 4 + 4}{4}$$

$$6 = \frac{4+4}{4} + 4 = \frac{4\sqrt{4}}{4} + 4$$

$$7 = \frac{44}{4} - 4 = \sqrt{4} + 4 + \frac{4}{4}$$

$$8 = 4 \cdot 4 - 4 - 4 = \frac{4(4+4)}{4}$$

$$9 = \frac{44}{4} - \sqrt{4} = 4\sqrt{4} + \frac{4}{4}$$

$$10 = 4 + 4 + 4 - \sqrt{4}$$

$$11 = \frac{4}{4} + \frac{4}{.4}$$

$$12 = \frac{4 \cdot 4}{\sqrt{4}} + 4 = 4 \cdot 4 - \sqrt{4} - \sqrt{4}$$

$$13 = \frac{44}{4} + \sqrt{4}$$

$$14 = 4 \cdot 4 - 4 + \sqrt{4} = 4 + 4 + 4 + \sqrt{4}$$

$$15 = \frac{44}{4} + 4 = \frac{\sqrt{4} + \sqrt{4} + \sqrt{4}}{.4}$$

$$16 = 4 \cdot 4 - 4 + 4 = \frac{4 \cdot 4 \cdot 4}{4}$$

$$17 = 4 \cdot 4 + \frac{4}{4}$$

$$18 = \frac{44}{\sqrt{4}} - 4 = 4 \cdot 4 + 4 - \sqrt{4}$$

$$19 = \frac{4 + \sqrt{4}}{.4} + 4$$

$$20 = 4 \cdot 4 + \sqrt{4} + \sqrt{4}$$

Topic: Determining Prime Numbers

Materials or Equipment Needed
A sheet of paper with a list of the numbers from 2 to 100 arranged as shown below.

Implementation of the Motivation Strategy
This motivational activity can be used to lead to a lesson and a class discussion of prime numbers and composite numbers. Prior to starting the lesson, it is a good idea to have an introductory discussion on prime numbers, prime factors, and composite numbers.

Give the students a list of the numbers from 2 through 100 as shown in table 3.2. Tell them to find all the prime numbers from 2 through 100. Remind them that the definition of a prime number[2] is one that has no factors other than itself and 1. Give them a few moments to try and to realize the difficulty of the task.

After attempting to find which numbers are prime by testing all the divisors of each number, the students should realize that there must be another way to find the prime numbers. Testing each number to find all possible divisors is very cumbersome.

TABLE 3.2 Sieve of Eratosthenes

	2	3	4	5	6	7	8	9	10
11	12	13	14	15	16	17	18	19	20
21	22	23	24	25	26	27	28	29	30
31	32	33	34	35	36	37	38	39	40
41	42	43	44	45	46	47	48	49	50
51	52	53	54	55	56	57	58	59	60
61	62	63	64	65	66	67	68	69	70
71	72	73	74	75	76	77	78	79	80
81	82	83	84	85	86	87	88	89	90
91	92	93	94	95	96	97	98	99	100

The Sieve of Eratosthenes is a historical curiosity that goes back to ancient Greece. None of Eratosthenes' (ca. 276–195 BC) works exist; the Sieve is referred to in the *Introduction to Arithmetic* by Nicomachus (ca. AD 60–120). Talk briefly about the meaning of a "sieve," designed to "strain out" all the non-primes, or composite numbers. Now show students that there is a device to sift out prime numbers that was discovered in ancient Greece thousands of years ago by a mathematician named Eratosthenes.

Have the students circle the first number, 2, and then cross out all multiples of 2 (in other words, the even numbers). Next, circle the first number not crossed out (in this case, 3) and cross out all multiples of 3. (Some of them, such as 6, 12, and 18, have already been eliminated as multiples of 2.) Then, circle the next number not crossed out (5) and cross out all multiples of 5. Continuing in this manner, circling the next number not crossed out and eliminating all multiples of that number, will leave the prime numbers.

2, 3, 5, 7, 11, 13, 17, 19, 23, 29, 31, 37, 41, 43, . . . 97.

These are the prime numbers. They have no factors except themselves and 1.

Looking at the Sieve, each of the prime numbers is circled, and all numbers of which that prime is a factor are crossed out. Thus, the only numbers remaining will be numbers with no factors other than themselves and one. This is the definition of a prime number.[3]

Topic: Algebraic Applications

Materials or Equipment Needed
No special materials are needed here.

Implementation of the Motivation Strategy
Beginning a lesson with a challenge will usually motivate students. The skill is to select challenges that will support the ensuing lesson rather than detracting from it. Take the following example:

Find five pairs of (rational) numbers whose product equals their sum.

Students are usually inclined to revert to an algebraic solution to a problem that resembles those they have had as exercises in an algebra class. Yet, presented with a simple arithmetic problem such as this one, most students will use trial-and-error methods with the hope of stumbling onto the correct answer. However, sometimes an algebraic solution can be superior.

Using the "intelligent guessing and testing" method should yield 2 and 2 as a pair of numbers satisfying the condition of having the same sum and product. From there on out it gets tougher since the rest of the numbers satisfying the required condition are fractions. This is where the usefulness of algebra sets in, especially to a rather frustrated group of students struggling to find other solutions after having found the first one rather easily—in short, a motivated population!

Let a and b represent the sought-after pair of numbers, such that $ab = a + b$. Then $a = \frac{b}{b-1}$. We can see how $a = 2, b = 2$ satisfies this equation. To find other pairs of numbers merely requires us to substitute values for b into the equation to get corresponding values for a. This can be seen in table 3.3.

The algebraic equation can be used to generate many other such pairs of numbers—a matter not so simple without algebra.

TABLE 3.3

b	$a = \frac{b}{b-1}$
3	$\frac{3}{2}$
4	$\frac{4}{3}$
5	$\frac{5}{4}$
6	$\frac{6}{5}$
−1	$\frac{1}{2}$
−2	$\frac{2}{3}$
−3	$\frac{3}{4}$

Topic: Introducing the Concept of π

Materials or Equipment Needed
Enough cardboard cores (cylinders) of toilet paper rolls for each group of students to have one.

Implementation of the Motivation Strategy
Holding the cardboard cylinder up for the class to see and each student group having one to look at more closely, ask students to conjecture: Which is greater, the circumference of the cylinder or the height of the cylinder (see figure 3.1)? Intuitively, it is expected that students will immediately say that the height is greater.

Besides using a string (or tape measure) to determine the answer to this challenge question, the teacher should ask the class how else this might be determined. Remind students of the definition of π, that is, the ratio of the circumference of a circle to its diameter. This should then immediately lead them to a formula for the circumference of a circle: $C = \pi d$. Thus they can see that the circumference is a bit more than three times the circle's diameter. Either with a ruler or simply marking off the length of the diameter and then marking it off three times along the height of the cylinder, students will be surprised to see how much greater the circumference is than the height—it is counter

FIGURE 3.1

intuitive: a real surprise. This will then lead into a lesson on the nature and properties of π.

Topic: Understanding the Value of π

Materials or Equipment Needed
A can with three tennis balls—one that is completely packed.

Implementation of the Motivation Strategy
The oddity of the teacher entering a classroom with a can of three tennis balls immediately stirs curiosity among the students. This curiosity in itself will stimulate the class's interest in the lesson. However, the teacher should begin by asking the class if the circumference of the can is longer or shorter than the height of the can—of course, without actually measuring it with a string or tape measure. Optically, students will generally guess that the height of the can is greater than the circumference.

If we have students use a string to do the comparison, they will find the difference so slight they may guess that the two lengths are equal. Now guide the students to analyze the situation. They should begin to realize that the height of the cylinder is exactly the same length as the sum of the three tennis ball diameters, which we can call $3d$. (See figure 3.2.)

FIGURE 3.2

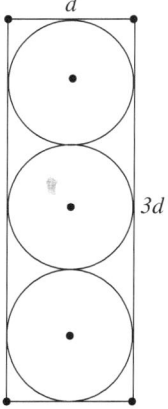

Through a variety of ways, the value of π has been established as the ratio of the circumference of a circle to its diameter. This means that the circumference of the circle at the top of the can is π times the diameter. We know that π is approximately equal to 3.14. Hence, the circumference of the top-of-the-can circle is about πd, or about $3.14d$, which is about .14 units longer than the height. Thus, the students now have a clear insight into what this constant ratio π is all about. The door is now open for a lesson on geometric measurements involving circles.

Topic: Introducing the Circumference of a Circle

Materials or Equipment Needed
Usual medium to present a problem—computer projector or chalkboard.

Implementation of the Motivation Strategy
Present the following problem to the students, but try to do it in an entertaining fashion:

> Consider a rope tied along the equator of the earth, circumscribing the entire earth sphere. Now lengthen this enormously-long rope by 1 meter. It is no longer tightly tied around the earth. If we lift this loose rope equally around the equator so that it is uniformly spaced above the equator, will a (real) mouse fit beneath the rope?

We are looking for the distance between the circumferences of these two circles (figure 3.3). Intuitively, the immediate reaction from students will be that a mouse could never fit under this rope. This could motivate students to want to find the circumference of the two circles. After this has been done in the traditional fashion, we will provide you with an unusual approach that exhibits a useful problem-solving technique.

Since the size of the circles is not given, suppose the small (inner) circle is extremely small, so small that it has a radius of length 0 and is thus reduced to a point. Then the distance between

FIGURE 3.3

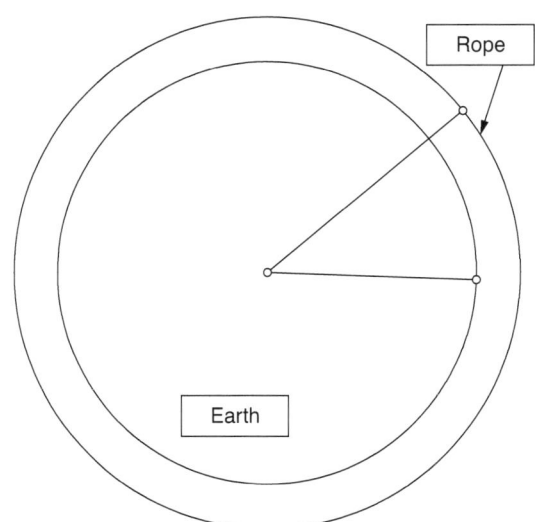

the circles is merely the radius of the larger circle. The circumference of this larger circle is $2\pi R = C + 1$, or $2\pi R = 0 + 1 = 1$ where C is the circumference of the earth (now for the sake of this problem reduced to 0) and $C + 1$ the length of the rope. The distance between the circles is $R = \frac{1}{2\pi} \cong 0.159$ meters, which would allow a mouse to comfortably fit beneath the rope.

This motivator is intended to demonstrate that even in geometry not everything is "intuitively obvious" and that there are geometric "facts" which not only seem to be wrong, but do not necessarily make sense without careful inspection. Most importantly it serves as an effective method to motivate students to a consideration of the circumference of a circle.

Topic: Finding the Sum of the Interior Angles of a Polygon

Materials or Equipment Needed
A medium to show the figures 3.4, 3.5, and 3.6.

Implementation of the Motivation Strategy
Assuming that the students know that the sum of the angles of a triangle is 180°, have each student draw any irregular hexagon.

| FIGURE 3.4 | FIGURE 3.5 | FIGURE 3.6 |

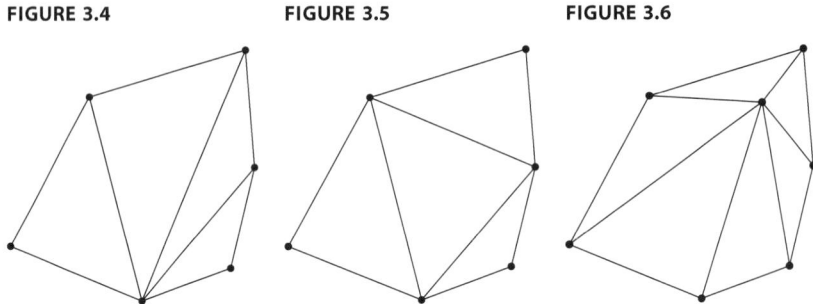

Then instruct the students to triangulate the hexagon—that is, have them draw lines in the hexagon to partition it into triangular regions. Ask the class to find the sum of the interior angles of the hexagon. Then ask them to extend this to a general n-gon. You may expect some drawings to look like one of the above figures.

This challenge—done with adequate patience on the part of the teacher to give students the time to come up with some creative triangulations—can lead directly to the formula that is to be taught in this lesson. If they arrived at the configuration of figures 3.4 or 3.5, then they will notice that the sum of the angles of the hexagon is the sum of the angles of four triangles or $4 \cdot 180° = 720°$.

If students arrived at the configuration of figure 3.6, then they will find that the sum of the interior angles of the hexagon is the same as six triangles minus the sum of the measures of the vertices at the interior point (360°). In this case the sum of the hexagon's angles is $(6 \cdot 180°) - (2 \cdot 180°) = 4 \cdot 180°$. In all three cases 180° was multiplied by $6 - 2 = 4$. With some other polygons the students should then realize that they derived the formula for the sum of the interior angles of an n-gon as $(n - 2)180°$. The initial challenge in this case has led to the goal of the lesson.

Topic: Proving Triangles Congruent

Materials or Equipment Needed
Usual medium to present a problem—computer projector or chalkboard.

Implementation of the Motivation Strategy

This strategy would be appropriate in the early lessons of a geometry course, after students have been introduced to proving triangles congruent. Here is a nice, challenging problem that essentially looks quite difficult but turns out to be very simple—if you "see it." The problem also lends itself nicely to a little story, but the essence of this motivation is "seeing" the triangles the students are to prove congruent in order to get the desired result. So we present the problem along with some enhancements.

Although the amazing geometric phenomenon we are about to present is attributed to Napoleon Bonaparte (1769–1821), some critics assert that the theorem was actually discovered by one of the many mathematicians[4] with whom Napoleon liked to interact.

Simply stated, we begin our exploration of this geometric novelty with a scalene triangle—that is, one that has all sides of different lengths. We then draw an equilateral triangle on each of the sides of this triangle. (See figure 3.7.)

Next, we will draw line segments joining the remote vertex of each equilateral triangle with the opposite vertex of the original triangle. (See figure 3.8.)

Ask the class to prove that the three dashed line segments in figure 3.8 are equal in length. Remember, this is true for a randomly selected triangle, which implies it is true for all triangles—that's

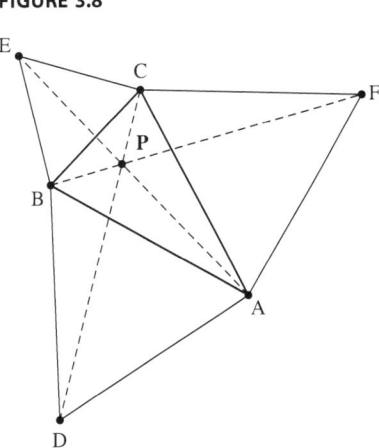

FIGURE 3.7

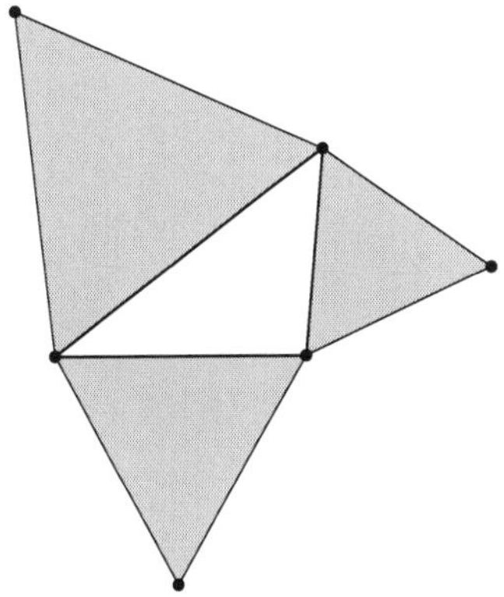

FIGURE 3.8

the amazing part of this relationship. The trick is to locate the congruent triangles. They are: $\triangle EBA \cong \triangle CBD$, and $\triangle ECA \cong \triangle BCF$, by the S.A.S. congruence theorem.

There are a host of amazing relationships in this configuration. You might mention some to the class and ask them to justify these properties and perhaps identify some others as well. For example, of all the infinitely many points in the original scalene triangle, the point of concurrency (not to be simply assumed, but proved) is the point from which the sum of the distances to the three vertices of the original triangle is the shortest.[5] That is, in figure 3.9, from the point P, the sum of the distances to the vertices A, B, and C (that is, $PA + PB + PC$) is a minimum. Also, the angles formed by the vertices of the original triangle at point P are equal. In figure 3.9, $m\angle APB = m\angle APC = m\angle BPC (= 120°)$. This point, P, is called the *Fermat point*, named after the French mathematician, Pierre de Fermat (1607–1665).

This configuration should serve as a motivational device in a number of ways and for various topics in the geometry curriculum.

FIGURE 3.9

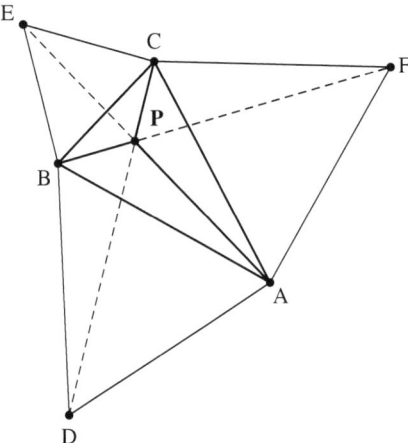

Topic: Introducing Geometric Series

Materials or Equipment Needed
Usual medium to present a problem—computer projector or chalkboard.

Implementation of the Motivation Strategy
Begin the lesson by presenting the following challenge to the class. "Would you rather have $100,000 per day for 31 days, or

1 cent the first day
2 cents the second day
4 cents the third day
8 cents the fourth day
16 cents the fifth day
And so on for 31 days?"

Experience shows that most students will opt for the first choice, namely, $100,000 dollars for each of 31 days, since that would amount to a rather large sum of $3,100,000. The task of adding the long list of cents amounts is also something they do not care to do. Determining this sum easily should be sufficient motivation for students to want to learn the topic of the day: find the sum of a geometric series. You might mention to them that they made a bad choice of money. (The sum of the cents option is $21,474,836.47!)

Notes

1 If students have learned that any number raised to the 0 power equals 1, you may provide an alternate solution to item 4 such as $(4 \div 4) - (4)^{(4-4)} = 0$.
2 The number 1 is neither a prime number nor a composite number.
3 An interesting sidelight of the lesson is to discuss Goldbach's Conjecture: Every even number greater than 2 can be expressed

as the sum of exactly two primes. Some examples include $8 = 3 + 5$, $10 = 3 + 7$, and $12 = 5 + 7$.

4 Any of the following might have "helped" Napoleon arrive at this relationship:

Jean-Victor Poncelet (1788–1867), who served as one of Napoleon's military engineers and who later became one of the founders of projective geometry; Gaspard Monge (1746–1818), a technical advisor to Napoleon who participated in the Egyptian campaign; Joseph-Louis Lagrange (1736–1813), a French mathematician; Lorenzo Mascheroni (1750–1800), who participated in the Italian campaign; Jean Baptiste Joseph de Fourier (1768–1830), who participated in the Egyptian campaign; Pierre-Simon Marquis de Laplace (1749–1827), who was Napoleon's teacher (1784–1785) and later served for six weeks as Napoleon's Minister of the Interior.

5 This is based on a triangle that has no angle greater than 120°. If the triangle has an angle greater than 120°, then the desired point is the vertex of the obtuse angle.

4

Entice the Class with a "Gee-whiz" Amazing Mathematical Result

One natural way to stimulate interest in mathematics among students is through the curiosity that nestles within us all. Such curiosity can be awakened through new ideas, paradoxes, uncertainties, or complex results. Here the teacher's talents come into play to find illustrations of easily understood situations that lead to unexpected results and leave the students intrigued (gee-whiz), resulting in a motivation to pursue the topic further.

Ideally, a problem or situation that is easily solved, that is short and doesn't distract from the intended lesson, and that leaves the students wondering "how and why," is ideal motivation for the topic to follow. So, for example, it is interesting to watch a student in a conundrum, when asked to determine if it's possible within an hour to gather apples which are one meter apart along a 100-meter stretch and place them individually into a stationary basket near the first apple. Almost all of the students guess that within an hour this should be no problem to accomplish and then are astonished to find out by calculation that the runner does not have the slightest chance of accomplishing this feat. The advantage of such an example is that it takes mathematics out of the abstract and brings it into the real-life experience of a student, increasing its relevance. There are many other examples of this sort that are counter-intuitive and at the same time appealing to the student's curiosity.

An effective motivation for probability is the famous birthday problem, shown in chapter 8 (page 120). This can be nicely demonstrated to a class by having the students quickly gather on slips of paper the birth dates of students in ten nearby classrooms, having about thirty students per room, and then checking to see if the expected seven out of ten classrooms have two people with the same birth date. This is a very powerful introduction to probability, as the results are quite different from what would be intuitively expected.

This and similar examples that lead to unexpected results, or a gee-whiz reaction, can entice students to pursue the mathematical background and thereby motivate them for the ensuing lesson. Naturally such examples are not always available, but for certain situations it is worth considering them, since the time invested is justified by the result.

Once the interest and attention of students, and thereby their cognitive reception, are captured, further study will be that much more effective. Once again we see that the students' self-generated enthusiasm through intrinsic motivation places them in a much more productive learning mode, spurred by this constant eagerness to satisfy the curiosity implanted by the teacher's earlier motivational activity. This is far more effective than the extrinsic motivation that comes from teacher praise, which often becomes tedious for teachers to use.

Topic: Introducing the Nature of Proof

Materials or Equipment Needed
Computer projector with prepared pictures of figures 4.1 through 4.5.

Implementation of the Motivation Strategy
This could be used as motivation for developing the notion of proof and securing its importance in the minds of the students. Your presentation of the following optical illusions may be as follows:

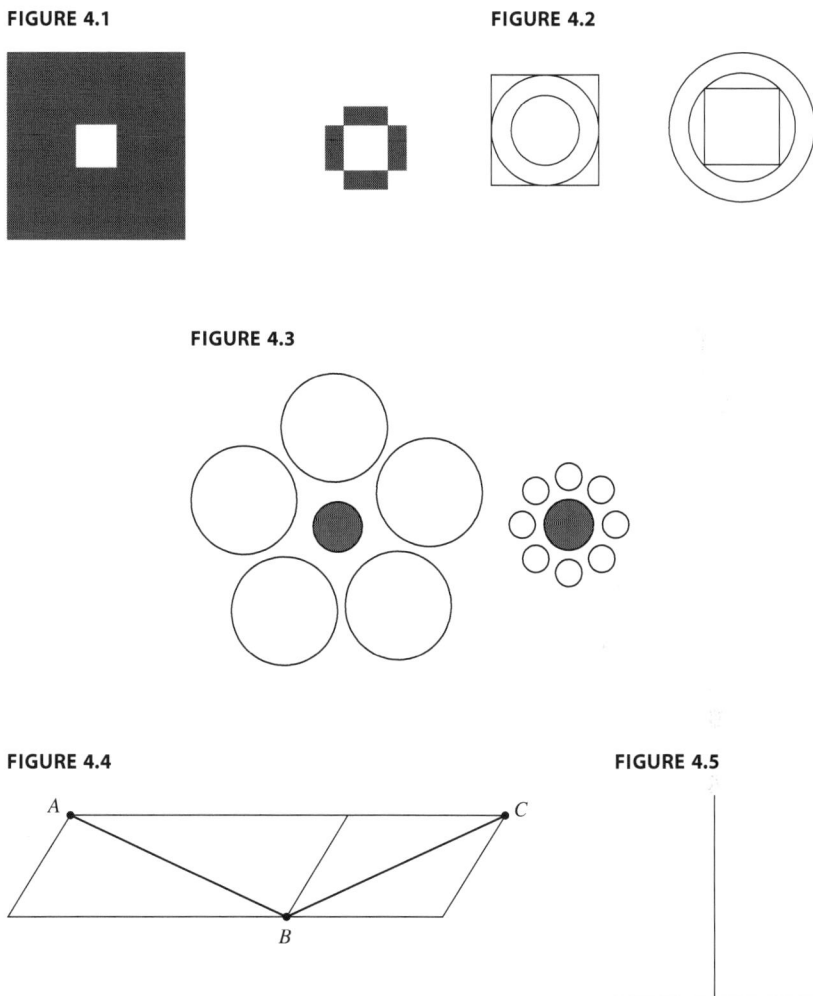

FIGURE 4.1

FIGURE 4.2

FIGURE 4.3

FIGURE 4.4

FIGURE 4.5

In geometry, what you see is not always what is true. For example, in figure 4.1 you see two white squares. Most observers would say that the left-side white square is a bit smaller than the white square on the right. This is incorrect. They are both the same size. Their context and presentation make for an optical illusion.

Another example to show you that your optical judgment may not be as accurate as you think is that of figure 4.2, where the circle inscribed in the square on the left appears to be smaller than the circle circumscribed around the square on the right. Again, this is not true, since they are the same size.

To further "upset" your sense of perception, consider the next few optical illusions. In figure 4.3, the center (black) circle on the left appears to be smaller than the center (black) circle on the right. Again, this is not so. They are both of equal size.

In figure 4.4, $\overline{AB}$ appears to be longer than $\overline{BC}$, but it isn't. They are the same length.

In figure 4.5, the horizontal line appears to be shorter than the vertical line, but, again, you have an optical illusion, since they are actually the same length.

This motivation should now have set the stage for whatever type of proof you have planned. Obviously, a geometric proof would be most fitting.

Topic: Thales' Theorem

Materials or Equipment Needed
A medium, such as a chalkboard or projection onto a board, to display the following figures.

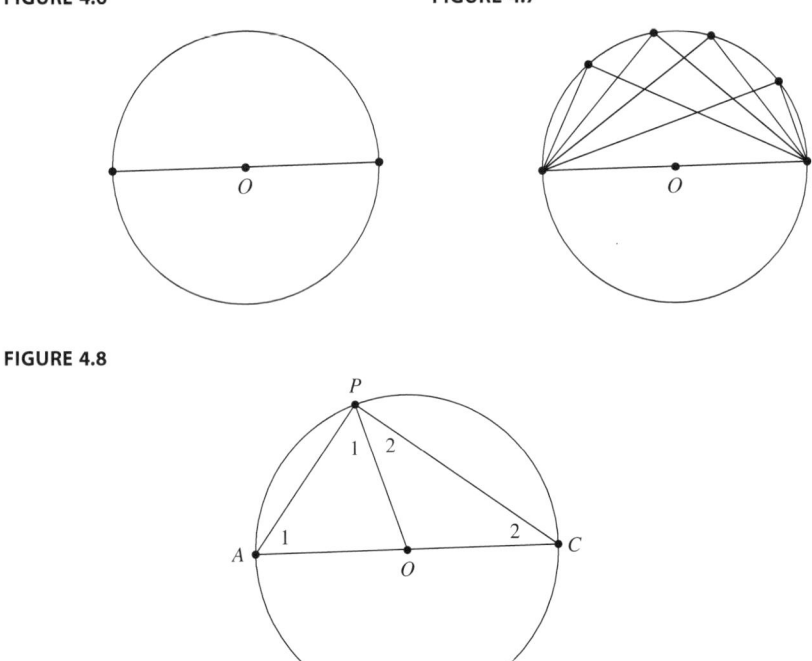

FIGURE 4.6

FIGURE 4.7

FIGURE 4.8

Implementation of the Motivation Strategy

Thales' Theorem states that: "Any angle inscribed in a semicircle is a right angle." This motivator leads to a proof of this theorem. Tell your class there will be a contest: Using only a compasses and straight edge, they are to draw or construct as many right angles as they can in two minutes. If they can construct more than you can in that given period, they will win a prize. Have them each take out a sheet of paper, ruler, and compasses and begin. (No protractors can be used except to check their work at the end, if you choose.)

The students should all know how to construct a right angle by constructing a perpendicular bisector of a given line segment. They should be able to construct a few in the given period, but you can easily win the contest by using Thales' Theorem.

While your students will begin to construct perpendiculars to form right angles, you construct a single circle and its diameter as shown in figure 4.6.

You can then draw as many right angles as you wish by simply inscribing them in the semi circle as in figure 4.7. You know that all the angles are 90 degrees because of Thales' Theorem.

The proof of this theorem is relatively simple and can be done as follows: Select any point, P, on the semicircle, and draw $\overline{AP}$ and $\overline{CP}$, as shown in figure 4.8. Because $\overline{OA}, \overline{OC}$, and $\overline{OP}$ are all radii of the same circle, we have two isosceles triangles, POA and POC. Thus their base angles are congruent. Using algebra, in triangle PAC,

$$\angle 1 + (\angle 1 + \angle 2) + \angle 2 = 180°$$
$$2 \cdot (\angle 1) + 2 \cdot (\angle 2) = 180°$$
$$\angle 1 + \angle 2 = 90°$$

and therefore the theorem is proved.

Of course, once the class has learned that an inscribed angle contains one-half the number of degrees as its intercepted arc, they know that all the angles are right angles since arc ADC is 180 degrees.

Topic: Introducing the Nature (or Importance) of Proof

Materials or Equipment Needed
A sheet with the list shown below.

$$3 = 2^0 + 2$$
$$5 = 2^1 + 3$$
$$7 = 2^2 + 3$$
$$9 = 2^2 + 5$$

Implementation of the Motivation Strategy
When a teacher embarks on the concept of doing a proof in mathematics, students often see this activity as another mathematical process they are required to learn without understanding the significance of proving that something is true for all cases. To motivate students, the teacher should demonstrate that you cannot just assume something is true because it appears that way. Begin the lesson by asking students if they believe that the following statement is true.

Every odd number greater than 1 can be expressed as the sum of a power of 2 and a prime number.

They should justify their response. Typically, students will try to see if this statement holds true for the first several cases.

In the short time allowed at the start of the lesson, this will probably suffice for the students to conclude that this is a true statement. This is where the "gee-whiz" factor appears, when you show them that this is true for all the odd numbers up to 125, but does not hold true for 127! This will truly shock them and will dramatically provide the justification for doing a proof before we can accept something as *always* true.

This is just one illustration of a number pattern that appears to lead to a general result—but does not. Let's consider the question of the French mathematician Alphonse de Polignac's (1817–1890) conjecture:

> "Every odd number greater than 1 can be expressed as the sum of a power of 2 and a prime number." (See table 4.1.)

TABLE 4.1

$$3 = 2^0 + 2$$
$$5 = 2^1 + 3$$
$$7 = 2^2 + 3$$
$$9 = 2^2 + 5$$
$$11 = 2^3 + 3$$
$$13 = 2^3 + 5$$
$$15 = 2^3 + 7$$
$$17 = 2^2 + 13$$
$$19 = 2^4 + 3$$
$$\ldots$$
$$51 = 2^5 + 19$$
$$\ldots$$
$$125 = 2^6 + 61$$
$$127 = ?$$
$$129 = 2^5 + 97$$
$$131 = 2^7 + 3$$

Perhaps you can ask students to find the next number that fails de Polignac's conjecture. They should remember, though, that when we say there is a pattern, we will need to make sure that it will hold true for *all* cases.

You might want to enrich your students with another conundrum. In 1849, Alphonse de Polignac proposed another conjecture that has not been proved or disproved to date. It is as follows:

> "There are infinitely many cases of two consecutive prime numbers greater than 2 with a difference of some even number n."

For example, suppose we let $n = 2$. There are an infinite number of consecutive prime number pairs whose difference is 2, such as (3, 5), (11, 13), (17, 19), etc. Note, we still have not established if this conjecture is true or false.

Topic: Considering Division by Zero

Materials or Equipment Needed
A medium, either a chalkboard or projection onto a board, to display the following "proof."

Implementation of the Motivation Strategy
One rule in mathematics that is not emphasized enough is that we cannot divide by zero. It is not always easy to find a situation that illustrates what happens if we mistakenly do divide by zero. The usual procedure is for the teacher to simply tell the class that they cannot divide by zero. This may not leave the class truly convinced as to why this is a forbidden division.

This motivator provides a simple situation for a class with basic algebraic skills. It will demonstrate what happens if we disobey this fundamental principle.

You now must dramatically tell the class that you have found a proof that 1 actually equals 2. They will laugh, and then you proceed to do the "proof," carefully going one step at a time on a whiteboard, chalkboard, or PowerPoint presentation. Have the class supply the reason for each step.

Assume (given)	$a = b$
(Multiply both sides of the equation by b.)	$ab = b^2$
(Subtract a^2 from both sides of the equation.)	$ab - a^2 = b^2 - a^2$
(Factor each side of the equation.)	$a(b - a) = (b + a)(b - a)$
(Divide both sides of the equation by $(b - a)$.)	$a = b + a$
(Substitute a for b on the right side of the equation.)	$a = a + a = 2a$
(Divide both sides of the equation by a.)	$1 = 2$

The students should experience some confusion at this point. Ask them to explain what we did that might have been wrong (if anything). They know (obviously) that something must be wrong, since 1 cannot equal 2. There is obviously a mathematical error. Somewhere we did something wrong. If we go back and examine each step, all appears well. So what is the problem?

In the fifth step, we divided both sides of our equation by the quantity $(b - a)$. Since we began by assuming that $a = b$ in step 1, we are, in essence, dividing by $(a - a)$ or 0. This should show the students that strange things can happen if we divide by zero. This is a motivating way to lead to a discussion of definitions in mathematics and why they are necessary. For example, one can prove that $-1 = +1$, if we allow that $\sqrt{ab} = \sqrt{a}\sqrt{b}$, for negative a and b.

Topic: The Introductory Lesson on Sample Space in Preparation for Probability

Materials or Equipment Needed
Six coins or chips: three of one kind and three of another.

Implementation of the Motivation Strategy
One of the least genuinely understood areas of school mathematics is probability. Students are usually interested, yet do not properly understand the concepts of "odds," "probability," "fair," and "sample space." This activity will provide motivation for a lesson on how and why to write out the complete sample space in a probability problem in which the outcomes are important. In the activity, they must use the sample space to decide if a game is fair or not. Writing out a sample space is an excellent technique to use when solving simple probability problems as it provides a visual display of the outcomes.

To begin, show the students a bag containing three chips. Tell them that two are red and one is black. They are to draw two chips from the bag without looking at them. If the chips are the same color, they win. If the chips are different colors, they lose. Is this a fair game?

The first term that must be clarified is "fair game." A "fair game" is one in which each player has an equal chance to win or lose. One example of a fair game is a coin toss; the chances of getting either heads or tails are the same.

Students, either in groups or as an entire class, should begin to experiment by trying the game. Actual chips and a bag can be used to have the class perform the experiment several times until they come to a decision. Intuitively, they might feel that the game is *not* fair. At that point, they should be asking if it is possible to determine what the outcome should be. This leads to a lesson that can demonstrate the power of mathematics by examining the sample space to determine if the game is fair or not. Mathematics enables them to "prove" the fairness of the game.

To show whether the game is fair or not, we can write out the sample space to illustrate the possibilities of the draw. There are three possibilities (R = red; B = black):

$$R_1B \quad R_2B_1 \quad R_1R_2$$

This is obviously not a fair game since only one of the three situations results in a win, namely R_1R_2.

Now the students should be asked to make the game fair. Have them consider placing exactly one more chip (either red or black) into the envelope. The majority of your students will suggest adding another black chip, thereby evening out the chips in the bag to two reds and two blacks. Again, the sample space should be used to determine possible outcomes. This time, there are six possible outcomes:

$$R_1B_1 \quad R_1B_2 \quad \mathbf{R_1R_2}$$
$$R_2B_1 \quad R_2B_2 \quad \mathbf{B_1B_2}$$

Again, this is *not* a fair game, since only two out of six possibilities result in a win.

Ask the class what else we might try to make this a fair game. Contrary to their intuitive guess, suppose we put in a third red chip. By now the students should suggest that they write out the sample space, consisting of six possibilities:

$$R_1B_1 \quad R_1R_2 \quad R_1R_3$$
$$R_2B_1 \quad R_2R_3 \quad R_3B_1$$

Aha! Now it is a fair game, since there are three out of six ways to win and three out of six ways to lose.

This result may surprise the students, since it goes against their expectations. It is an excellent way to motivate them to write out the sample space when confronted by a difficult probability problem. It provides a visual display of the outcomes.

Topic: Introduction to the Concept of Area, or Looking Beyond the Expected

Materials or Equipment Needed
A computer projector with dynamic geometry software, such as Geometer's Sketchpad, would be best, but a conventional chalkboard would suffice.

Implementation of the Motivation Strategy
Here is a problem that looks very simple but is not. It has baffled entire high school mathematics departments! Yet once the solution is shown, it becomes quite simple. The result is that your students will be disappointed in not having seen the solution right from the start. Try it without looking at the second diagram, which would give away the solution. This problem should get a "gee-whiz" reaction from your students and make them receptive to the rest of the lesson, which is aimed at showing students how to "think out of the box" in approaching a mathematical problem.

In figure 4.9, point E lies on $\overline{AB}$ and point C lies on $\overline{FG}$.

FIGURE 4.9

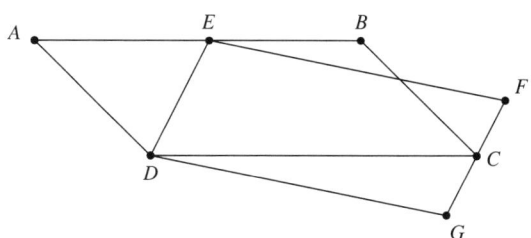

The area of parallelogram $ABCD = 20$ square units. Find the area of parallelogram $EFGD$.

The solution is not one that would occur to many students at first thought, as they have been focusing on congruence and similarity for much of the geometry course. Yet, the problem can be readily solved using only the tools found in a high school geometry course. Begin by drawing $\overline{EC}$ as in figure 4.10.

Since triangle EDC and parallelogram $ABCD$ share a common base ($\overline{DC}$) and a common altitude (a perpendicular from E to $\overline{DC}$), the area of triangle EDC is equal to one-half the area of parallelogram $ABCD$.

Similarly, since triangle EDC and parallelogram $EFGD$ share the same base ($\overline{ED}$), and the same altitude to that base (a perpendicular from C to $\overline{ED}$), the area of triangle EDC equals one-half the area of parallelogram $EFGD$.

Now, since the area of parallelogram $ABCD$ and the area of parallelogram $EFGD$ are both equal to twice the area of triangle EDC, the areas of the two parallelograms must be equal. Thus, the area of parallelogram $EFGD$ equals 20 square units.

Although the solution method that we have just shown is not often used, it is effective and efficient. Nevertheless, this problem can be solved quite elegantly by solving a simpler analogous problem (without loss of generality). Recall that the originally given conditions were that the two parallelograms had to have a common vertex (D), and a vertex of one had to be on the side of the other as shown with points E and C. Now, let us suppose that C coincided with G, and E coincided with A. This satisfies the given condition of the original problem and makes the two parallelograms coincide. Thus the area of parallelogram $EFGD = 20$ square units.

FIGURE 4.10

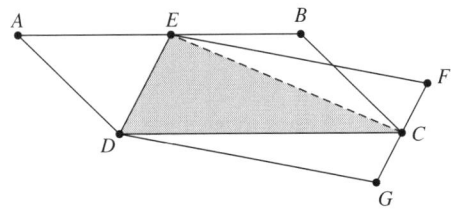

We could also look at this last solution as one of using extremes. That is, we might consider point E on $\overline{AB}$, yet placed at an extreme, such as on point A. Similarly, we could place C on G and satisfy all the conditions of the original problem. Thus the problem is trivial, in that the two parallelograms overlap. This point is one of the more neglected techniques for solving problems. It ought to be emphasized now.

Remember how difficult your students perceived the problem at the start? This gee-whiz reaction should serve as a good motivator for the remainder of the lesson as long as the teacher keeps to the focus set up in this initial activity.

Topic: Introduction to the Area of a Circle or to Finding Areas of Similar Figures

Materials or Equipment Needed
A computer projector with dynamic geometry software, such as Geometer's Sketchpad would be best, but a conventional chalkboard would suffice.

Implementation of the Motivation Strategy
When a teacher presents the following to a class, the result is met with disbelief or a gee-whiz response.

> Given 5 concentric circles, as shown in figure 4.11, with the inner circle of radius 1, and each successing circle having a radius 1 unit larger than the previous one. Which is larger, the outer ring (the area between the larger two circles) or the shaded region?

FIGURE 4.11

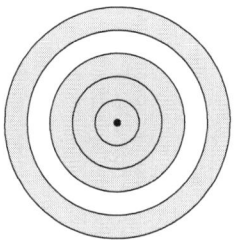

Intuitively, the center shaded region appears to be greater than the outer ring. Students will be shocked to learn and calculate that the two areas are equal. This will lead nicely (and enthusiastically) into the lesson so that this conjecture can be verified.

Topic: Infinite Geometric Series

Materials or Equipment Needed
The usual display medium will suffice here.

Implementation of the Motivation Strategy
To dramatize the topic of the ensuing lesson, infinite series, begin with a physical demonstration as follows: Stand against the wall directly opposite the door to your classroom. Explain to the students that you wish to leave the classroom. However, your plan for leaving is to continuously walk to a point that is halfway to the door. Then walk to that point. This leaves one-half the distance from the wall to the door still to be covered. Walk to a point one-half the remaining distance. Again, one-half the distance (one-fourth of the original distance) to the door remains. Continue by walking one half of this remaining distance. This leaves one-half the distance (or one-eighth of the original distance). Ask the class if you will ever reach the door and leave the room.

The students will, of course, be certain that you will reach the door—and, physically, they are correct! But is that mathematically correct? At this point, you can discuss the series $\frac{1}{2}+\frac{1}{4}+\frac{1}{8}+\frac{1}{16}+\cdots$, which represents the amount of distance *remaining* between the wall and the door as you walk. Does this distance ever reach 1? Or is there always one-half of *some* distance remaining? Is there a mathematical method for resolving this apparent conflict? This leads to a lesson on the infinite geometric series with ratio less than 1.

The infinite geometric series with a common ratio less than one is a mathematical puzzlement to many students. Obviously, you can *physically* reach the doorway. However, mathematically, the sum of the sequence $\frac{1}{2}, \frac{1}{4}, \frac{1}{8}, \frac{1}{16}, \cdots$ is never ending. The sum of the series is given by the formula $S = \frac{a}{1-r}$, where a is the first term of the series and r is the common ratio of the geometric series. Here, $a = \frac{1}{2}$, and $r = \frac{1}{2}$. Therefore, here we have $S = \frac{\frac{1}{2}}{\frac{1}{2}} = 1$.

Thus the limit of the sum of the terms in this series is 1 (the total distance from the wall to the door). Students should understand that this represents the *limit* of the sum of the terms in the series, and mathematically you will never reach the door. You can create a sense of amusement by continuing out the door.

5

Indicate the Usefulness of a Topic

In recent years it has become fashionable to show the usefulness in everyday life of the subjects being taught in school. This has been particularly true in mathematics, where challenges to the need for teaching the subject constantly arise. Even students who achieve high grades in mathematics tend to question its usefulness. Although logical thinking that develops during the study of mathematics should be enough to demonstrate its usefulness, this aspect of the subject still comes into question despite the fact that many students have not even selected a career path.

As is the case with other motivational devices presented in this book, the surprise factor plays a role when a particular topic's usefulness leads into the lesson. Thus, indicating the usefulness of a topic can provide a fine motivation for the students to want to learn the topic to be presented. There are many examples of mathematics being applied in a wide variety of fields that can appeal to a number of different age groups.

Where possible, teachers should select mathematical applications as lesson motivators that come from the students' set of experiences. They could range from percent considerations such as price increases on previously discounted items, or examples from sports statistics, best routes for students to take from school to home, and so on.

It is also possible that family members of various students in the class have interesting occupations that lend themselves to mathematical applications. In some circumstances inviting these family members to the class to present a mathematical application from their field could be quite energizing for the entire class.

Once again, the teacher should be cautious not to allow this mathematical application to dominate the lesson, bearing in mind that it is merely a motivator.

Topic: Introduction to Proportions

Materials or Equipment Needed
The usual display medium: chalkboard or computer projector.

Implementation of the Motivation Strategy
Present the following challenge problem to the class. Experience shows that most students cannot get a grasp on this problem since there are so many variables.

If a apples cost d dollars, then at the same rate, what is the cost in cents of b apples?

This seemingly simple problem has proved to be quite challenging for unsuspecting students. Because it appears simple and because few students in the class typically get the right answer in a reasonable amount of time, the problem peaks the students' interest and serves nicely to motivate students for a lesson on proportions. It will clearly demonstrate the usefulness of the topic of proportions.

Students try to use some sort of reasoning and perhaps even try to substitute numbers for the letters so as to make the problem less abstract. However, it should be rather easy to convince them of the power of a proportion to solve this problem; thus making this a fine lead into a discussion of proportions.

To solve the problem with proportions, they merely have to set up the following proportion: $\frac{a \text{ apples}}{b \text{ apples}} = \frac{100d \text{ cents}}{x \text{ cents}}$, then solving for x, gives us: $\frac{100bd}{a}$. The simplicity of the solution with the help of proportional thinking should motivate the class to pursue other such problems and practice their proportions.

Topic: Applying Algebra

Materials or Equipment Needed
The usual display medium: chalkboard or computer projector.

Implementation of the Motivation Strategy
Most mathematics teaching standards expect students to be able to represent and analyze mathematical situations using the symbols of algebra. This motivator involves the students in a "mind reading" activity that challenges them to explain how and why the instructions lead to a particular result. They then learn that they must use their algebraic skills to prove the results. This motivator provides an excellent activity to introduce a lesson on algebraic proof or justification.

Begin by having students write any 3-digit number (consisting of three different digits). Then have them form all the possible two-digit arrangements that can be made from the three digits of the original number (there should be six such permutations or arrangements of the digits). Have them add all six of the two-digit numbers. Next, they are asked to divide this sum by the sum of the digits of their original number. All the students will have a result of 22. Ask them why this happens.

At first, students will be "amazed" that everyone's answer is 22. Their curiosity should be aroused to ask why this happens, regardless of the three digits they started with. They should ask you to show why this always works. This leads to a lesson on algebraic proofs or justifications.

The proof involves representing two- and three-digit numbers algebraically and following the directions carefully. A three-digit number can be written as $100h + 10t + u$ where h, t, and u are the three digits. The sum of the digits is $(h + t + u)$.

Their work should look something like the following:
Write a 3-digit number (with 3 different digits): $100h + 10t + u$
Form all six two-digit permutations:

$$10h + t$$
$$10t + h$$
$$10u + h$$
$$10h + u$$
$$10t + u$$
$$10u + t$$

Add these six numbers to get: $22h + 22t + 22u = 22(h + t + u)$
Divide by the sum of the digits of their original number:

$$22(h + t + u) \div (h + t + u) = 22$$

Students will then see the justification for why 22 always remains the answer.

Topic: Introduction to Similar Triangles

Materials or Equipment Needed
The usual display medium: chalkboard or computer projector.

Implementation of the Motivation Strategy
This motivator leads to a series of lessons on similar triangles and how to make use of the proportionality of their corresponding sides. It develops this feature for making indirect measurements for objects that cannot be measured directly. Tell the story of how Thales, the famous ancient Greek mathematician and engineer, was asked to measure the height of the pyramids in ancient Egypt, an inaccessible height. He found a clever technique to accomplish this task. You could ask the class how he might have been able to do this.

FIGURE 5.1

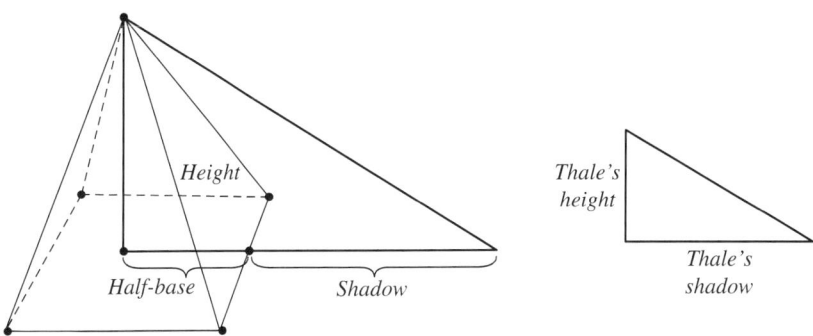

Some students will, of course, suggest using a ladder. This is not feasible because the height of the pyramid would create the need for a very long ladder. Second, the ladder would not be able to stand in the center of the pyramid to get a vertical measurement. Discuss some of their ideas before revealing what Thales did.

Thales made use of the triangle formed by three points—the top of the pyramid, the external endpoint of its shadow, and the base of the pyramid's altitude. He created a pair of similar triangles using his own height and shadow, and compared it to the length of the shadow of the pyramid. (See figure 5.1.)

Since the triangles are similar, he made use of the following proportion:

$$\frac{\text{Thales' height}}{\text{Pyramid's height}} = \frac{\text{Thales' shadow}}{\text{Pyramid's shadow} + \text{half base}}$$

This process is sometimes referred to as "shadow reckoning."

Topic: Introducing Modular Arithmetic

Materials or Equipment Needed
The usual display medium: chalkboard or computer projector.

FIGURE 5.2

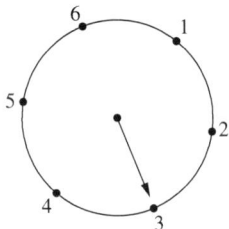

Implementation of the Motivation Strategy
Present the following to the class with the task of determining why this can make sense.

$$3 + 7 = 14$$

This is clearly something that conflicts with what students already "know" to be correct. The initial reaction will be that the example is wrong. Assure them that it is quite correct. Ask if they have ever seen a gas meter. (See figure 5.2.) The result will easily lead into a lesson on modular arithmetic, where the students will be guided to see that when you start at 3 and move around the gas meter 7 units we get to 4, but since we moved around once to get there we must make that 1, 4 (mod 6).

Topic: Introduction to the Concurrency of the Angle Bisectors of a Triangle

Materials or Equipment Needed
Any medium to show the following problem, preferably a dynamic geometry computer software program such as Geometer's Sketchpad or GeoGebra.

Implementation of the Motivation Strategy
Begin the class with a presentation of a problem that will motivate students to learn the relationship of the angle bisectors of a triangle—namely, that they are concurrent. The problem that

FIGURE 5.3

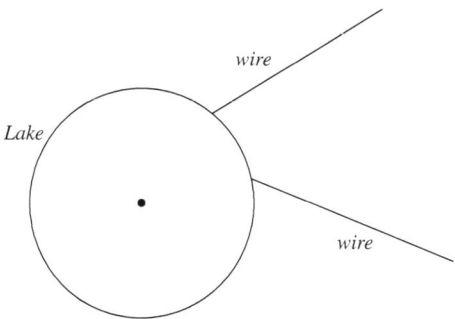

the class is to grapple with is as follows: We have two wires in a field next to a lake. Each ends at the shoreline of the lake. If they were extended, they would intersect in the lake. The problem is to place a third wire between these two wires so that this third wire would bisect the angle formed by the other two wires—had they been extended into the lake. (See figure 5.3.)

This problem will show the usefulness of knowing that the angle bisectors of a triangle are concurrent.

Begin by drawing any line through the two wires and then the bisectors of the angles thus formed, as in figure 5.4.

Students should now realize that the desired angle bisector—the one bisecting the inaccessible angle in the lake—must contain the point, P, at which these two angle bisectors intersect.

Repeating this procedure for another line that intersects the two given wires gives us the point Q. (See figure 5.5.) This time the bisector of the third angle of the triangle thus formed would have to also contain point Q, the point of intersection of the angle bisectors of the second triangle formed.

Students should now realize that having points P and Q both on the bisector of the inaccessible angle in the lake will determine the desired line, shown in figure 5.6. This problem ought to have served to demonstrate the usefulness of the concurrency of the angle bisectors of a triangle.

FIGURE 5.4

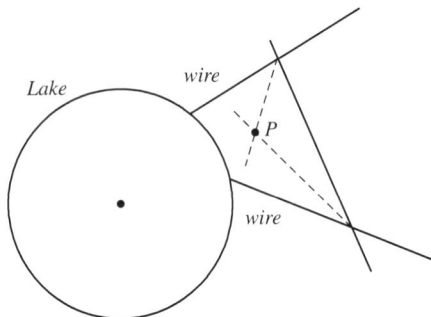

FIGURE 5.5

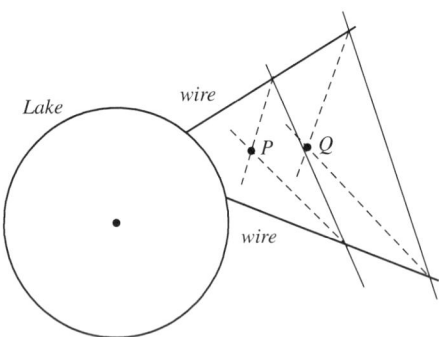

FIGURE 5.6

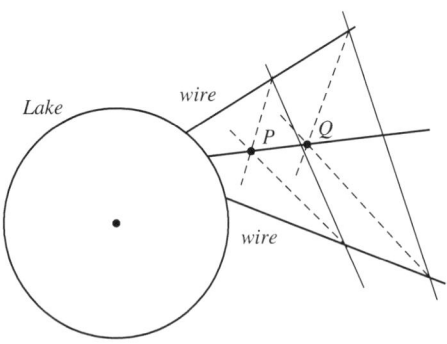

Topic: Determining the Volume of a Right Circular Cylinder

Materials or Equipment Needed
For each student or group of students: two sheets of 8½" × 11" paper, cellophane tape, and a supply of beans.

Implementation of the Motivation Strategy

Students should learn how to find the volume of a cylinder. As they work through the typical unit on volume, they usually begin with cubic and rectangular solids, all with sides of integral side-lengths, figures that hold small 1 × 1 × 1 cubes. These cubes can then be counted to determine the volume, and the appropriate formulas are developed. When working with a cylinder, however, these cubes cannot completely fill the interior of the figure and another method needs to be developed. This activity should motivate students to want to find the formula for the volume of a cylinder, which is the ensuing lesson.

Give each group of students two sheets of 8½" by 11" paper and some cellophane tape. Also give each group a supply of beans. Have each group make a cylinder by rolling the paper along the 11" side and taping it to form the cylinder. Have them make a second cylinder by rolling the paper along the 8½" side and taping it. The students should be asked to find out which cylinder holds the greater volume.

Since the students have not been exposed to the formula for the volume of a cylinder, they must devise another way to solve this problem. Students usually have the initial reaction that they both hold the same amount, since they were made from the same size sheet of paper. After some experimenting, they should arrive at the conclusion that they can fill each cylinder with the beans, pour them out and, by counting, determine the greater volume.

The method the students use to find which cylinder has the greater volume is a valid one. However, it does involve a great amount of careful filling of the cylinders and then counting the number of beans contained in each. Also, while this method does answer the question of which one has the *greater* volume, it does not enable us to find out the actual volume of each cylinder. For that they need the formula, $V = \pi r^2 \cdot h$, where h is the height of the cylinder and r is the radius of the circular bottom (or top). This is the lesson for the day.

After the lesson has been presented, we can resolve the original students' problem as follows:

> Cylinder #1: Height = 11", base radius is found by using the circumference formula. The circumference is 8.5".

$C = 2\pi r$, then $r = 1.35$. Students can now find the volume of the first cylinder:

$V = \pi r^2 h$, then $V = \pi \cdot (1.35)^2 \cdot 11 = 62.98$ cubic inches.

Cylinder #2: Height = 8.5″, base radius is found by using the circumference formula. The circumference is 11″. $C = 2\pi r$, then $r = 1.75$. Here students will again use the formula:

$V = \pi r^2 h$, so that $V = \pi \cdot (1.75)^2 \cdot 8.5 = 81.78$ cubic inches.

Rather than using beans and having to tediously count them, applying the formula is a far easier and more exact method for finding the volume of a cylinder. The students' initial motivation of finding the usefulness of the topic should be clear.

Topic: Introduction to Probability— Expected Outcomes

Materials or Equipment Needed
For each student, an answer sheet of paper with space for 20 true-false responses.

Implementation of the Motivation Strategy
This motivator makes an excellent introduction to probability. It makes the students a partner in the outcomes and opens the door to more study of probability. It challenges them to decide whether or not it is worthwhile to guess on a test.

Give each student a sheet of paper and have them number from 1 through 20 for a True-False test. Tell the class that this is an unusual test—one for which they will not have the questions. They are simply to guess whether each question is answered true or false. If they think the item is true, they should indicate it with a T; if they think it is false, indicate it with an F. When they are finished, tell the class what the correct "answers" are and have them score their papers.

We shall use the following "correct answers" arrived at by coin tosses. If the toss showed heads, the item was considered true; tails was considered false.

1. F	5. F	9. T	13. T	17. T
2. F	6. F	10. F	14. F	18. F
3. T	7. T	11. T	15. T	19. F
4. F	8. F	12. F	16. F	20. F

The students will probably laugh at the situation and use some "system" for putting T or F on each question. Those with some vague ideas about probability may even write T for all the answers, and assume they will have at least 50 percent correct. When done, they may ask why there were more Fs (13) than Ts (7). Intuitively, they would expect 10 of each. This leads to a lesson on empirical versus experimental probability.

Have the students then combine all of their Ts and Fs. See if the combined result approaches 50 percent. Be sure the class understands that, as the number of cases increases, the experimental probability will more closely approach the expected probability. Discuss the concept of the sample space. In this case, the first two answers could have been T-F, T-T, F-T, or F-F.

The probability of getting two consecutive false answers is only 1 out of 4, or $\frac{1}{4}$. Does it make sense to guess? What is the probability of getting 100 percent on this test by guessing? (Answer: $\left[\frac{1}{2}\right]^{20}$.) Students ought to then be asked if they can think of a situation where the expected outcome is 100 percent and one where the expected outcome is 0 percent.

Topic: Introducing the Product of the Segments of Two Intersecting Chords of a Circle

Materials or Equipment Needed

Any medium to show the following problem, preferably a dynamic geometry computer software program such as Geometer's Sketchpad or GeoGebra.

Implementation of the Motivation Strategy

Begin by telling the students that you had a problem recently. A plate broke and you needed to get a replacement. The largest piece of the broken plate was less than a semicircle, and therefore you had the problem of determining what the diameter of the plate was. This is shown in figure 5.7, where the dashed line represents the cracked portion.

Following the lines drawn in figure 5.8, we can measure the segments: $\overline{AE}$, $\overline{BE}$, and $\overline{CE}$, as 6″, 6″, and 3″, respectively. This leads to the "cross-chords" theorem, which gives us $AE \cdot BE = CE \cdot DE$, or $6 \cdot 6 = 3x$, and $x = 12$. Therefore, the diameter of the plate was $3″ + 12″ = 15″$.

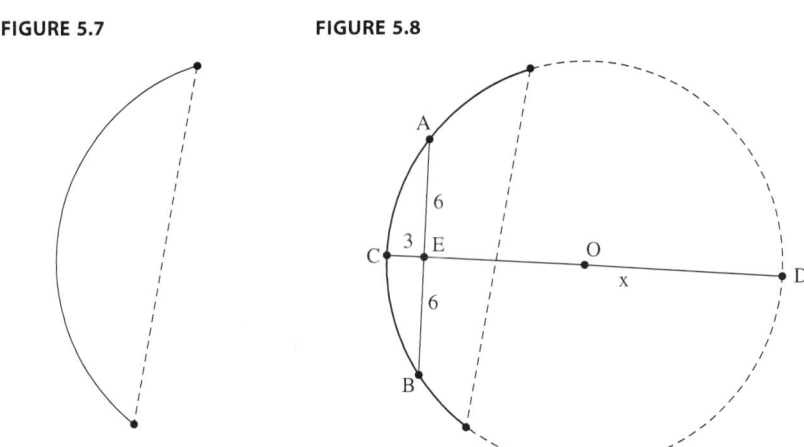

FIGURE 5.7 **FIGURE 5.8**

Topic: Introduction to the Concurrency of the Altitudes of a Triangle

Materials or Equipment Needed

Any medium to show the following problem, preferably a dynamic geometry computer software program such as Geometer's Sketchpad or GeoGebra.

Implementation of the Motivation Strategy

The circumcenter of a triangle is the point at which the perpendicular bisectors of the sides are concurrent. This motivator leads

FIGURE 5.9

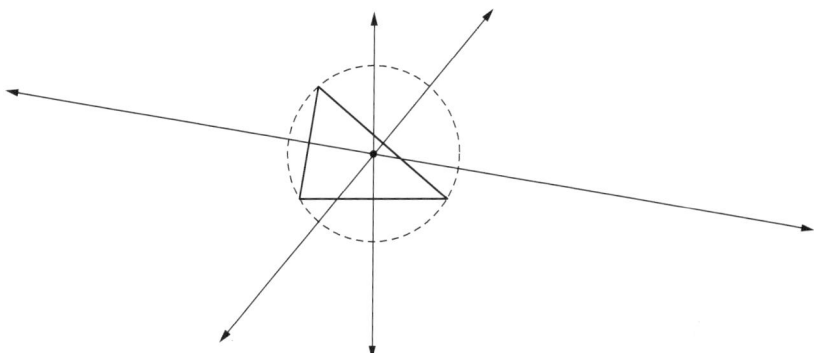

to a lesson on the *circumcenter, incenter,* and *orthocenter* of a triangle. The students should be motivated by placing the following location problem for them to consider. The Wagner Food Company is planning to build a new food distribution center that will be equally distant from each of three of its best customers.[1] Determine how they might find the location to build this center.

Students should quickly discover that the location of the three clients determines a triangle. Thus, what they are looking for is a point that is equally distant from the three vertices. This should lead them to ask how to find this point. A lesson on the circumcenter should follow immediately.

The lesson will then cover the location of this point, where the three perpendicular bisectors of the sides of a triangle are concurrent (figure 5.9). This point of concurrency, called the circumcenter of the triangle, can be outside the triangle (an obtuse triangle), on a side of the triangle (a right triangle), or inside the triangle (an acute triangle), and is equidistant from the three vertices of the triangle. The Wagner Food Company should, therefore, build their distribution center at the circumcenter of the triangle formed by the three locations. A bit later the class might want to know the location of the point where the sum of the distances to the three customers is a minimum. (See page 50.)

Note

1 Note, the three customers are not at three collinear points.

6

Use Recreational Mathematics

Most people enjoy playing games. This includes solving puzzles, riddles, and other kinds of mathematical problems. In recent years the number puzzle Sudoku has captured the fascination of both children and adults. Properly selected, a recreational math activity can also be useful to teachers by providing a lead-in to a lesson. Not only is this motivational, it also should give the students a feeling of achievement.

Students should be able to appreciate the fun and recreational aspect of the activity and thereby develop an interest in the ensuing lesson. Teachers should be cautious in selecting the activity to avoid detracting from the lesson that follows, bearing in mind that the activity should motivate students' interest in the theme of the lesson.

Some recreational mathematics can be an integral part of the lesson; others can be used "just for fun." Recreational mathematics is full of puzzles, games, paradoxes, and curiosities. In addition to being selected for their specific motivational power, these devices must be challenging, yet brief and enjoyable. Although they are sometimes impractical, they are fun, increase interest, stimulate intellectual curiosity, and allow the development of both mathematical techniques and concepts. A student should be able to achieve success in doing the "recreation" without too much effort in order for this technique to be effective. Many motivational recreations can be justified by a simple use of algebra. For instance, most "think of a number" tricks lend themselves readily to algebraic solutions. Here is just one example:

Step 1: Think of a number.	x
Step 2: Double your number.	$2x$
Step 3: Add 8.	$2x + 8$
Step 4: Subtract 2.	$2x + 6$
Step 5: Divide by 2.	$x + 3$
Step 6: Subtract your original number.	3

Your result is 3.

This simple example of algebraic "magic" will amaze your students, and lead them into a lesson on algebraic representation and proof.

Ideal recreations are those that appear difficult and yet are surprisingly simple to resolve. Such solutions can often require thinking "outside the box." Illustrations can be found in many forms of mathematical thinking and problem processing. In selecting these motivational devices, teachers should be mindful not to make them so easy for the intended population that they become silly, or so difficult that they are beyond the reach of most students. In assessing the appropriateness of these recreational motivational devices, one needs to consider that a student's emotional development is also influenced by factors that occur within the classroom and can change with the various stimuli provided by the teacher. Thus teachers' assessments of students should be flexible to anticipate changes throughout the course. As a student develops his or her intellectual capacity, so, too, will the appreciation, understanding, and processing of a puzzle evolve with time.

Unfortunately too many adults still harbor a fear of or dislike of mathematics largely because their teachers did not take into account the importance of motivating instruction—especially using recreational techniques. This sort of motivational device is an excellent way of not only introducing the lesson, but also demonstrating a light-hearted aspect of mathematics that can have a lasting effect on students.

Reasoning problems and mathematical games can demonstrate the fun that mathematics can offer. Such activities often win over the uninitiated and serve as much more than mere motivation for the forthcoming lesson.

Topic: Identifying Factors of Numbers

Materials or Equipment Needed
An appropriate medium to display the mathematics.

Implementation of the Motivation Strategy
Begin the lesson by asking students to find all the proper factors of 220 (except for 220 itself) and then finding the sum of these factors. They should then do the same with the number 284. If they did this correctly, they should have come up with an unusual result. These two numbers can be considered "friendly numbers"!

What could possibly make two numbers friendly? Your students' first reaction might be that these numbers are friendly to them. Remind them that we are talking here about numbers that are "friendly" to each other. Mathematicians have decided that two numbers are considered friendly (or, as often used in the more sophisticated literature, "amicable") if the sum of the proper divisors of the first number equals the second *and* the sum of the proper divisors of the second number equals the first.

Sounds complicated? Have your students now consider the smallest pair of friendly numbers: 220 and 284. They should have gotten the following, which should show the recreational aspect of mathematics.

The divisors of **220** are 1, 2, 4, 5, 10, 11, 20, 22, 44, 55, and 110.
Their sum is $1 + 2 + 4 + 5 + 10 + 11 + 20 + 22 + 44 + 55 + 110 = $ **284.**

The divisors of **284** are 1, 2, 4, 71, and 142, and their sum is $1 + 2 + 4 + 71 + 142 = $ **220.**

This shows that the two numbers are friendly numbers.

You might now want to embark on the broader concept of determining factors or take this recreational theme one step further.

A second pair of friendly numbers discovered by Pierre de Fermat (1601–1665) is: 17,296 and 18,416.

Here $17{,}296 = 2^4 \cdot 23 \cdot 47$, and $18{,}416 = 2^4 \cdot 1151$.

The sum of the factors of 17,296 is

$$1 + 2 + 4 + 8 + 16 + 23 + 46 + 47 + 92 + 94 + 184 \\ + 188 + 368 + 376 + 752 + 1081 + 2162 \\ + 4324 + 8648 = \mathbf{18{,}416}$$

The sum of the factors of 18,416 is $1 + 2 + 4 + 8 + 16 + 1151 + 2302 + 4604 + 9208 = \mathbf{17{,}296}$

Here are a few more pairs of friendly numbers:

1,184 and 1,210
2,620 and 2,924
5,020 and 5,564
6,232 and 6,368
10,744 and 10,856
9,363,584 and 9,437,056
111,448,537,712 and 118,853,793,424

Your students might want to verify the above pairs' "friendliness," although this may be quite time consuming.

Topic: Understanding Percents

Materials or Equipment Needed
The usual display medium.

Implementation of the Motivation Strategy
You can present this any way that would make it appear recreational for the intended student audience. Applying it to some local concern would make it even more recreational. It also has to be presented in an appropriate tone. We will demonstrate it in a general way merely as a model.

Suppose you had a job where you received a 10 percent raise. Because business was falling off, the boss was soon forced to give you a 10 percent cut in salary. Will you be back to your starting salary? The answer is a resounding (and very surprising) NO!

This little story is quite disconcerting, since one would expect that with the same percent increase and decrease you should be back from where you started. This is intuitive thinking, but wrong. Students may want to convince themselves of this by choosing a specific amount of money and trying to follow the instructions.

Begin with $100. Calculate a 10 percent increase on the $100 to get $110. Now take a 10 percent decrease of this $110 to get $99—$1 less than the beginning amount.

They may wonder whether the result would have been different if they first calculated the 10 percent decrease and then the 10 percent increase. Using the same $100 basis, first calculate a 10 percent decrease to get $90. Then the 10 percent increase yields $99, the same as before. So order apparently makes no difference.

This will lead the teacher into a discussion of percents and percentages with a now-motivated class.

Topic: Reinforce Some Logical Thought in Mathematical Work

Materials or Equipment Needed
The usual mathematics display medium.

Implementation of the Motivation Strategy
Problem solving is not only done to solve the problem at hand; it is also provided to present various types of problems and, perhaps more importantly, various procedures for solution. It is from the types of solutions that students really learn problem solving, since one of the most useful techniques in approaching a problem to be solved is to ask yourself: "Have I ever encountered such a problem before?" With this in mind, a problem with a very useful "lesson" is presented here. Do not let your students be deterred by the relatively lengthy reading required to get through the problem. They will be delighted with its ultimate simplicity.

Two trains, serving the Chicago to New York route, a distance of 800 miles, start towards each other, at the same time (along the same tracks). One train is traveling uniformly at 60 miles per hour, and the other at 40 miles per hour. At the same time, a bumblebee begins to fly from the front of one of the trains, at a speed of 80 miles per hour towards the oncoming train. After touching the front of this second train, the bumblebee reverses direction and flies towards the first train (still at the same speed of 80 miles per hour). The bumblebee continues this back and forth flying until the two trains collide, crushing the bumblebee. How many miles did the bumblebee fly before its demise?

Students will be naturally drawn to find the individual distances that the bumblebee traveled. An immediate reaction by many students is to set up an equation based on the relationship: "rate times time equals distance". However, this back and forth path is rather difficult to determine, requiring considerable calculation. Just the notion of having to do this will cause frustration among the students. Do not allow this frustration to set in. Even if they were able to determine each part of the bumblebee's flight, it is still very difficult to solve the problem in this way.

A much more elegant approach would be to solve a simpler analogous problem (one might also say we are looking at the problem from a different point of view). We seek to find the distance the bumblebee traveled. If we knew the time the bumblebee traveled, we could determine the bumblebee's distance because we already know the speed of the bumblebee. Again, have your students realize that having two parts of the equation will provide the third part. So having the time and the speed will yield the distance traveled, albeit in various directions.

The time the bumblebee traveled can be easily calculated, since it traveled the entire time the two trains were traveling towards each other (until they collided). To determine the time, t, the trains traveled, we set up an equation as follows: The distance of the first train is $60t$ and the distance of the second train is $40t$. The total distance the two trains traveled is

800 miles. Therefore, $60t + 40t = 800$, and $t = 8$ hours, which is also the time the bumblebee traveled. We can now find the distance the bumblebee traveled, using the relationship, which gives us $(8)(80) = 640$ miles.

It is important to stress for students how to avoid falling into the trap of trying to do what the problem calls for directly. Sometimes a more circuitous method is much more efficient. Lots can be learned from this solution. It must be emphasized to your class. You see, dramatic solutions are often more useful that traditional solutions, since it gives students an opportunity to observe "thinking outside of the box."

Topic: Rationalize the Denominator of a Fraction

Materials or Equipment Needed
The usual mathematics display medium.

Implementation of the Motivation Strategy

Begin with what would appear to be a recreational activity in that the students will see that this appears to be a "trick question." Consider the following series and ask students to find the sum.

$$\frac{1}{\sqrt{1}+\sqrt{2}} + \frac{1}{\sqrt{2}+\sqrt{3}} + \frac{1}{\sqrt{3}+\sqrt{4}} + \cdots$$
$$+ \frac{1}{\sqrt{2009}+\sqrt{2010}} + \frac{1}{\sqrt{2010}+\sqrt{2011}}$$

Students often find rationalizing a denominator merely an exercise without much purpose. Naturally they are given applications that show a need for this technique, but somehow these applications usually do not convince students of the usefulness of the procedure. There are applications (somewhat dramatic) that drive home the usefulness argument quite nicely, such as the one just posed.

Students are taught that they cannot do much with a fraction where the denominator is irrational and so must seek to change it to an equivalent fraction with a rational denominator. To do this they know to multiply the fraction by 1 so as not to change its value. Yet, the form that 1 should take is a fraction with the conjugate of the current denominator in both the numerator and denominator.

The general term of this series may be written as $\frac{1}{\sqrt{k}+\sqrt{k+1}}$. To rationalize the denominator of this fraction, multiply it by 1 in the form of $\frac{\sqrt{k}-\sqrt{k+1}}{\sqrt{k}-\sqrt{k+1}}$ to get: $\frac{1}{\sqrt{k}+\sqrt{k+1}} \cdot \frac{\sqrt{k}-\sqrt{k+1}}{\sqrt{k}-\sqrt{k+1}} = \frac{\sqrt{k}-\sqrt{k+1}}{-1}$.

That is, we have found $\frac{1}{\sqrt{k}+\sqrt{k+1}} = \sqrt{k+1} - \sqrt{k}$.

This huge simplification will allow students to rewrite the series as:

$$\left(\sqrt{2}-\sqrt{1}\right)+\left(\sqrt{3}-\sqrt{2}\right)+\left(\sqrt{4}-\sqrt{3}\right)+\cdots$$
$$+\left(\sqrt{2009}-\sqrt{2010}\right)+\left(\sqrt{2010}-\sqrt{2011}\right),$$

which then eliminates many of the terms by subtraction and then becomes simply:

$$\sqrt{2011} - 1 \approx 44.844175 - 1 = 43.844175.$$

Students can readily see how rationalizing the denominator is not just an exercise without use. It is useful, and here we have had a prime example.

Topic: Applications of Algebra Explaining Arithmetic Peculiarities

Materials or Equipment Needed
The usual mathematics display medium.

Implementation of the Motivation Strategy
To motivate students for a lesson on algebraic applications, you might entertain them with the following request:

Ask students to reduce to lowest terms the following fractions:

$$\frac{16}{64}, \frac{19}{95}, \frac{26}{65}, \frac{49}{98}.$$

After they have reduced to lowest terms each of the fractions in the usual manner, ask why they didn't simply do it in the following way:

$$\frac{1\cancel{6}}{\cancel{6}4} = \frac{1}{4}$$

$$\frac{1\cancel{9}}{\cancel{9}5} = \frac{1}{5}$$

$$\frac{2\cancel{6}}{\cancel{6}5} = \frac{2}{5}$$

$$\frac{4\cancel{9}}{\cancel{9}8} = \frac{4}{8} = \frac{1}{2}$$

At this point your students should be somewhat amazed. Their first reaction is probably to ask if this can be done to any fraction composed of two-digit numbers of this sort. Challenge your students to find another fraction (comprised of two-digit numbers) where this type of cancellation will work. Students might cite $\frac{55}{55} = \frac{5}{5} = 1$ as an illustration of this type of cancellation. Indicate to them that although this will hold true for all multiples of eleven yielding two-digit results, it is trivial, and our concern will be only with proper fractions (those whose value is less than one).

Students should now be motivated to see if these are the only such fractions that allow this silly reduction. This is then when the motivated students embark on "explaining" this situation. The four fractions above, in fact, are the only ones (composed of two-digit numbers) where this type of cancellation will hold true.

Have students consider the fraction $\frac{10x+a}{10a+y}$.

The above four cancellations were such that when canceling the a's the fraction was equal to $\frac{x}{y}$.

Therefore, $\frac{10x+a}{10a+y} = \frac{x}{y}$.

This yields:
$$y(10x + a) = x(10a + y)$$
$$10xy + ay = 10ax + xy$$
$$9xy + ay = 10ax$$

And so
$$y = \frac{10ax}{9x+a}.$$

At this point, have students inspect this equation. They should realize that it is necessary that x, y, and a are integers, since they were digits in the numerator and denominator of a fraction. It is now their task to find the values of a and x for which y will also be integral.

To avoid a lot of algebraic manipulation, you might have students set up a chart which will generate values of y from $y = \frac{10ax}{9x+a}$. Remind them that x, y, and a must be single-digit integers. Table 6.1 shows a portion of the table they will construct. Notice that the cases where $x = a$ are excluded, since then $\frac{x}{a} = 1$.

TABLE 6.1

$x \backslash a$	1	2	3	4	5	6	...	9
1		$\frac{20}{11}$	$\frac{30}{12}$	$\frac{40}{13}$	$\frac{50}{14}$	$\frac{60}{15}=4$		$\frac{90}{18}=5$
2	$\frac{20}{19}$		$\frac{60}{21}$	$\frac{80}{22}$	$\frac{100}{23}$	$\frac{120}{24}=5$		
3	$\frac{30}{28}$	$\frac{60}{29}$		$\frac{120}{31}$	$\frac{150}{32}$	$\frac{180}{33}$		
4								$\frac{360}{45}=8$
⋮								
9								

The portion of the chart pictured above already generated two of the four integral values of y; that is, when $x = 1$, $a = 6$, then $y = 4$, and when $x = 2$, $a = 6$, then $y = 5$. These values yield the fractions $\frac{16}{64}$ and $\frac{26}{65}$, respectively. The remaining two integral values of y will be obtained when $x = 1$ and $a = 9$, yielding $y = 5$, and when $x = 4$ and $a = 9$, yielding $y = 8$. These yield the fractions $\frac{19}{95}$ and $\frac{49}{98}$, respectively. This should convince students that there are only four such fractions composed of two-digit numbers.

Topic: Applications of Algebraic Counterintuitive Peculiarities

Materials or Equipment Needed
Each student should bring 12 pennies to class.

Implementation of the Motivation Strategy
When we teach a lesson on algebraic applications, we often leave students questioning the real need for these applications. To counteract this negative attitude, begin this class with the following recreational example. This lovely little problem will show your students how some clever reasoning along with *algebraic knowledge* of the most elementary kind will help solve a seemingly "impossibly difficult" problem.

Here is the problem, which can be worked on by students individually or in groups, whichever fits the class more appropriately.

> You are seated at a table in a dark room. On the table there are 12 pennies, 5 of which are heads up and 7 of which are tails up. (You know where the coins are, so you can move or flip any coin, but because it is dark you will not know if the coin you are touching was originally heads up or tails up.) You are to separate the coins into two piles (possibly flipping some of them) so that when the lights are turned on there will be an equal number of heads in each pile.

Your first reaction is "you must be kidding! How can anyone do this task without seeing which coins are heads or tails up?" This is where a clever (yet incredibly simple) use of algebra will be the key to the solution.

After students have had ample time to get a bit frustrated, the teacher may lead the discussion through pointed questioning. Have them begin by separating the coins into two piles, of 5 and 7 coins each. Then flip over the coins in the smaller pile. Now both piles will have the same number of heads! That's all heads! Your students will think this is magic. How did this happen? Well, this is where algebra helps in understanding what was actually done.

When you separate the coins in the dark room, h heads will end up in the 7-coin pile. Then the other pile, the 5-coin pile, will have $5 - h$ heads. To get the number of tails in the 5-coin pile, we subtract the number of heads $(5 - h)$ from the total number of coins in the pile, 5, to get: $5 - (5 - h) = h$ tails. (See table 6.2.)

TABLE 6.2

5-Coin Pile	7-Coin Pile
$5 - h$ heads	h heads
$5 - (5 - h)$ tails $= h$ tails	

When you flip all the coins in the smaller pile (the 5-coin pile), the $(5 - h)$ heads become tails and the h tails become heads. Now each pile contains h heads! (See table 6.3.)

TABLE 6.3 The piles after flipping the coins in the smaller pile

5-Coin Pile	7-Coin Pile
$5 - h$ tails	h heads
h heads	

This absolutely surprising result will show them how the simplest algebra can explain a very complicated reasoning exercise and hopefully motivate the students to place greater value on algebraic applications.

Topic: Introduction to Divisibility Rules, Especially Divisibility by 11

Materials or Equipment Needed
The usual mathematics display medium.

Implementation of the Motivation Strategy

Most people know that to multiply by 10 we merely have to tag a zero onto the number. For example, when multiplying 78 by 10, we get 780. You can motivate students to consider other multiplications, such as by 11, and do them very quickly. We suggest showing students the following mental method of multiplying by 11 before discussing divisibility by 11.

The simpler a mathematical "trick" is, the more attractive it tends to be. Here is a very nifty way to multiply by 11. This one always gets a rise out of the unsuspecting mathematics-phobic person, because it is so simple that it is even easier than doing it on a calculator!

The rule is very simple:

> To multiply a two-digit number by 11 just add the two digits and place this sum between the two digits.

Let's try using this technique. Suppose you wish to multiply 45 by 11. According to the rule, add 4 and 5 and place their sum, 9, between the 4 and 5 to get 495.

This can get a bit more difficult if the sum is a two-digit number. What do we do in that case? We no longer have a single digit to place between the two original digits. So, if the sum of the two digits is greater than 9, we place the units digit between the two digits of the number being multiplied by 11 and "carry" the tens digit to be added to the hundreds digit of the multiplicand.[1] Let's try it with 78 · 11.

7 + 8 = 15. We place the 5 between the 7 and 8, and add the 1 to the 7, to get [7 + 1][5][8] or 858.

You may legitimately ask if the rule also holds when 11 is multiplied by a number of more than two digits.

Let's go right for a larger number such as 12,345 and multiply it by 11.

Here we begin at the units digit and add every pair of digits going to the left.

$$1[1 + 2][2 + 3][3 + 4][4 + 5]5 = 135{,}795.$$

If the sum of two digits is greater than 9, then use the procedure described before: place the units digit of that sum appropriately and carry the tens digit. We will do one of these here.

Multiply 456,789 by 11.

We carry the process step by step:

4[4 + 5][5 + 6][6 + 7][7 + 8][8 + 9]9
4[4 + 5][5 + 6][6 + 7][7 + 8][17]9
4[4 + 5][5 + 6][6 + 7][7 + 8 + 1][7]9
4[4 + 5][5 + 6][6 + 7][16][7]9
4[4 + 5][5 + 6][6 + 7 + 1][6][7]9
4[4 + 5][5 + 6][14][6][7]9
4[4 + 5][5 + 6 + 1][4][6][7]9
4[4 + 5][12][4][6][7]9
4[4 + 5 + 1][2][4][6][7]9
4[10][2][4][6][7]9
[4 + 1][0][2][4][6][7]9
[5][0][2][4][6][7]9
5,024,679

This rule for multiplying by 11 is usually so well-received by students that they are not only interested in the number 11—and hence ready to discuss its divisibility rule—but will likely share this newly acquired knowledge with family and friends.

Try to convince your students that at the oddest times the issue can come up of a number being divisible by 11. If you have a calculator at hand, the problem is easily solved. But that is not always the case. Besides, there is such a clever "rule" for testing for divisibility by 11 that it is worth showing students just for its charm.

The rule is quite simple: **If the difference of the sums of the alternate digits is divisible by 11, then the original number is also divisible by 11.** Sounds a bit complicated, but really isn't. Have your students take this rule a piece at a time. The sums of the alternate digits means you begin at one end of the number taking the first, third, fifth, etc. digits and add them. Then add the remaining (even placed) digits. Subtract the two sums and inspect for divisibility by 11.

It is probably best shown to your students by example. We shall test 768,614 for divisibility by 11. Sums of the alternate digits are: $7 + 8 + 1 = 16$ and $6 + 6 + 4 = 16$. The difference of these two sums, $16 - 16 = 0$, which is divisible by 11.[2]

Another example might be helpful to firm up your student's understanding. To determine if 918,082 is divisible by 11, find the sums of the alternate digits: $9 + 8 + 8 = 25$ and $1 + 0 + 2 = 3$. Their difference is $25 - 3 = 22$, which is divisible by 11, and so the number 918,082 is divisible by 11.[3]

Now just let your students practice with this rule. They will like it better with more practice, and they will love showing it to their family and friends.

This should lead students to want to have divisibility rules for other numbers—a topic well worth the time to motivate students towards mathematics in general!

Topic: Application of Algebraic Solutions to Digit Problems

Materials or Equipment Needed
The usual mathematics display medium.

Implementation of the Motivation Strategy
Students are often suspicious about review or practice with algebraic problems, so it would be refreshing to begin a lesson with a recreational activity that will motivate the students to discover the secret that causes amazement. This trick is fun, and will show how we can analyze a seemingly baffling result through simple algebra. Begin by asking the students to select any three-digit number with no two like digits (omitting the

zero). Then, have them make five other numbers with these same digits.[4] Suppose a student selected the number 473, then the list of all numbers formed from these digits is:

$$473$$
$$437$$
$$347$$
$$374$$
$$743$$
$$734$$

We can get the sum of these numbers, 3,108, faster than they can even write the numbers. How can this be done so fast? All we actually have to do is get the sum of the digits of the original number (here: we get $4 + 7 + 3 = 14$) and then multiply 14 by 222 to get 3,108, which is the required sum. Why 222? Let us inspect some of the many quirks of number properties, using simple algebra:

Consider the number $abc = 100a + 10b + c$, where $a, b, c \in \{1, 2, 3, \ldots, 9\}$.

The sum of the digits is $a + b + c$. We now represent all of the six numbers of these digits on our list:

$$100a + 10b + c$$
$$100a + 10c + b$$
$$100b + 10a + c$$
$$100b + 10c + a$$
$$100c + 10a + b$$
$$100c + 10b + a$$

This equals:

$$100(2a + 2b + 2c) + 10(2a + 2b + 2c) + 1(2a + 2b + 2c)$$
$$= 200(a + b + c) + 20(a + b + c) + 2(a + b + c)$$
$$= 222(a + b + c),$$ which is 222 times the sum of the digits.

If you really want to be slick and beat the students to the answer, then you might want to have the following chart (table 6.4) on a small piece of paper for easy reference:

TABLE 6.4

Digit sums	6	7	8	9	10	11	12	13	14	15
Six-Number Sum	1,332	1,554	1,776	1,998	2,220	2,442	2,664	2,886	3,108	3,330
Digit Sum	16	17	18	19	20	21	22	23	24	
Six-Number Sum	3,552	3,774	3,996	4,218	4,440	4,662	4,884	5,106	5,328	

That's all we need to avoid the actual addition. Students should be motivated to delve into other such digit problems. It is incumbent upon the teacher to select motivating applications of this skill.

Notes

1. The multiplicand is the number that is multiplied by another number, the multiplier. In arithmetic, the multiplicand and the multiplier are interchangeable, depending on how the problem is stated, because the result is the same if the two are reversed—for example, 2 × 3 and 3 × 2. Therefore, 2 × 3 means "add 2 three times," whereas 3 × 2 means "add 3 two times."
2. Remember $\frac{0}{11} = 0$.
3. For the interested student, here is a brief discussion about why this rule works as it does. Consider the number ab, cde, whose value can be expressed as
$N = 10^4 a + 10^3 b + 10^2 c + 10 d + e = (11-1)^4 a + (11-1)^3 b + (11-1)^2 c + (11-1)d + e$
$= [11M + (-1)^4]a + [11M + (-1)^3]b + [11M + (-1)^2]c + [11 + (-1)]d + e$
$= 11M[a + b + c + d] + a - b + c - d + e$, which implies that divisibility by 11 of N depends on the divisibility of: $a - b + c - d + e = (a + c + e) - (b + d)$, the difference of the sums of the alternate digits.
Note: $11M$ refers to a multiple of 11.
4. If you have been exposed to permutations before, you will recognize that these six numbers are the *only* ways to write a number with three different digits, which is why we ask for three *distinct* digits in the first place!

7

Tell a Pertinent Story

In exploring the history of mathematics, one encounters stories and anecdotes which are both amusing and interesting and consequently worth retelling. This is largely attractive because some are truly original and humorous, while others show the unusual thinking of famous mathematicians and how they approached and conquered problems, despite unusual personalities, and often through very elementary ways.

Through such stories, mathematical equations, formulas, and symbols take on a different meaning than the mere visual. Consequently students have another form of recollection of these concepts. It also makes students aware of the sometimes slow and deliberate fashions in which concepts have been developed over the years, and not as an instant result on paper or the chalkboard—as they are too frequently presented in school. In some cases, mathematicians labor their entire lives and involve others to reach a solution or to understand a mathematical concept properly.

To this we must add that most people—and in particular children—love a good story. Who can't remember having heard a story told, curious to see how it unfolds and eager to know the conclusion? In order to create this emotional response in the listener, one naturally needs to be a talented storyteller who plans in advance what he would like to achieve by having told the story. This is the role of the teacher, who should secure the precision of the story, gather the appropriate details, and take into account the age and interests of the student audience. Of course the timing of the story must be appropriate as its intention is to provide motivation for the ensuing lesson.

It is important to stress that only a teacher who is comfortable with the theme of a story, plans appropriate pauses, and tells the story with enthusiasm, will make this story an effective pedagogical tool. A poorly told story could, in fact, have a negative result—the opposite of what is intended. Another concern in telling the story is to avoid rushing through the story quickly, just to get to the conclusion which leads to the mathematical topic of the day. Such rapid and perfunctory storytelling cannot only be counterproductive, but also a pure waste of time. Perhaps one way of gauging a properly told story is to compare it to a properly-told joke, where the timing, pacing, and coloring all contribute to its success.

The age of the students is also an important consideration in selecting and telling a story. Eleven- and twelve-year-old students are still entertained with an emotional storytelling style, while older students sometimes find it silly and therefore are distracted from the story's contents. For older students, a more content-oriented story will be more engaging; yet the teacher should exercise caution to avoid making it boring or too dry.

A teacher who succeeds in selecting the right story at the right time, presented in the right tone with proper length, has an excellent chance to awaken the students' interest in preparation for the ensuing lesson.

As simple as it appears to tell a good story, that's how difficult it can be to select an appropriate story with proper preparation and self-reflection as to whether one can tell a story properly. If not, despite the story's appropriateness for the intended lesson, an alternative strategy might be a better approach.

This chapter presents some stories from the history of mathematics and other related experiences that can be used as a motivational device for an effective instructional experience.

There are many interesting stories about historical figures and episodes in the history of mathematics. There are also many resources for teachers to capture appropriate stories, such as *In Mathematical Circles*, by Howard W. Eves, or *Men of Mathematics*, by Eric Temple Bell.

Topic: Introducing Divisibility Rules

Materials or Equipment Needed
The usual mathematics classroom equipment.

Implementation of the Motivation Strategy
Begin the lesson with the following "story": Tell the class that you were at a restaurant with two other people the other day, and when it came time to pay the bill you wanted to add a tip and be sure that the total amount would be divisible by 3. Your friends wanted to try different numbers and use a calculator to check whether the sum would be divisible by 3. You told the group that you could determine this by simply looking at the number and doing an instant calculation. The class will then be curious and motivated to determine how this can be done. This is where you introduce the notion that the sum of the bill will only be divisible by 3 if the sum of the digits is also divisible by 3. Hence, if the sum was $74.33, then the check could not be divided exactly by 3, since $7 + 4 + 3 + 3 = 17$, which is not divisible by 3. Whereas, if the check were $74.34, then the check could be divided by 3 exactly, because the sum of the digits is 18, which is divisible by 3. Once this has been established, the class ought to also consider the rule for divisibility by 9, since the proof for divisibility by 9 will encompass the rule for divisibility by 3.

Topic: Introduction to the Value of π

Materials or Equipment Needed
Prepared visuals showing the Hebrew letters as described below; a piece of string and a ruler as well as some circular object for each student.

Implementation of the Motivation Strategy

To appreciate this revelation of π, you need to know that many books on the history of mathematics state that in its earliest manifestation in history, namely the Bible (Old Testament), the value of π was given as 3. Yet recent "detective work" shows otherwise.[1]

Telling this story in a motivating fashion will generate much more interest in the value of π than if the concept is presented traditionally. Students always relish the notion that a hidden code can reveal long-lost secrets. Such is the case with the common interpretation of the value of π in the Bible. There are two places in the Bible where the same sentence appears, identical in every way except for one word, spelled differently in the two citations. The description of a pool or fountain in King Solomon's temple is referred to in the passages that may be found in 1 Kings 7:23 and 2 Chronicles 4:2, and read as follows:

> And he made the molten sea of ten cubits from brim to brim, round in compass, and the height thereof was five cubits; and *a line* of thirty cubits did compass it round about.

The circular structure described here is said to have a circumference of 30 cubits and a diameter of 10 cubits. (A cubit is the length from a person's fingertip to his elbow.) From this we notice that the Bible has $\pi = \frac{30}{10} = 3$.

This is obviously a very primitive approximation of π. A late 18th century Rabbi, Elijah of Vilna (Poland), one of the great modern biblical scholars, who earned the title "Gaon of Vilna" (meaning brilliance of Vilna), came up with a remarkable discovery, one that could make most history of mathematics books faulty if they say that the Bible approximated the value of π as 3. Elijah of Vilna noticed that the Hebrew word of "line measure" was written differently in each of the two Biblical passages mentioned above.

In 1 Kings 7:23 it was written as קוה, whereas in 2 Chronicles 4:2 it was written as קו. Elijah applied the biblical analysis technique (still used today) called gematria, where the Hebrew letters are given their appropriate numerical values according to their sequence in the Hebrew alphabet, to the two spellings of the word for "line measure" and found the following.

The letter values are: ק = 100, ו = 6, and ה = 5. Therefore, the spelling for "line measure" in 1 Kings 7:23 is קוה = 5 + 6 + 100 = 111, while in 2 Chronicles 4:2 the spelling is קו = 6 + 100 = 106. He then took the ratio of these two values: $\frac{111}{106} = 1.0472$ (to four decimal places), which he considered the necessary correction factor, for when it is multiplied by 3, which is believed to be the value of π stated in the Bible, one gets 3.1416, which is π correct to four decimal places! "Wow!!!" is a usual reaction. Such accuracy is quite astonishing for ancient times.[2] To support this notion, have students take a string to measure the circumference and diameter of several circular objects and find their quotient. They will most likely not get near this four-place accuracy. Moreover, to really push the point of the high degree of accuracy of four decimal places, chances are if you took the average of all the students' π measurements, you still wouldn't get to four-place accuracy.

Topic: Introduction to Prime Numbers

Materials or Equipment Needed
The usual mathematics classroom equipment.

Implementation of the Motivation Strategy
Begin the lesson with the story of how there are some problems posed in mathematics to which no solution has ever been found. Sometimes the problem is stated as a guess, or conjecture, without proof, and therefore cannot be accepted as *always true*. One such case centers on Christian Goldbach (1690–1764), a German mathematician, who in a June 7, 1742 letter to Leonhard Euler (1707–1783), posed the following statement, which to this day has yet to be verified. *Goldbach's Conjecture* is as follows:

> Every even number greater than 2 can be expressed as the sum of two prime numbers.

You might ask students to begin with the following list of even numbers and their prime-number sum and then continue it to convince themselves that it continues on—apparently—indefinitely. (See table 7.1.)

TABLE 7.1

Even numbers greater than 2	Sum of two prime numbers
4	2 + 2
6	3 + 3
8	3 + 5
10	3 + 7
12	5 + 7
14	7 + 7
16	5 + 11
18	7 + 11
20	7 + 13
...	...
48	19 + 29
...	...
100	3 + 97

Again, there have been substantial attempts by famous mathematicians to prove the conjecture. In 1855, A. Desboves verified Goldbach's Conjecture for up to 10,000 places. Yet in 1894, the famous German mathematician Georg Cantor (1845–1918) (regressing a bit) showed that the conjecture was true for all even numbers up to 1,000; it was then shown by N. Pipping in 1940 to be true for all even numbers up to 100,000. By 1964, with the aid of a computer, it was extended to 33,000,000; in 1965 this was extended to 100,000,000; and then in 1980 to 200,000,000. Then, in 1998, the German mathematician Jörg Richstein showed that Goldbach's conjecture was true for all even numbers up to 400 trillion. On February 16, 2008, Oliveira e Silva extended this to 1.1 quintillion ($1.1 \times 10^{18} = 1,100,000,000,000,000,000$)! Prize money of \$1,000,000 has been offered for a proof of this conjecture. To date, this has not been claimed.

Topic: Finding the Sum of an Arithmetic Series

Materials or Equipment Needed
A photograph of Carl Friedrich Gauss (1777–1855) would be nice to exhibit—available from various websites.

Implementation of the Motivation Strategy
While a student in elementary school in the eighteenth century, the young Carl Friedrich Gauss (1777–1855), who later went on to become one of the greatest mathematicians in history, had as his teacher Mr. Buettner, who one day wanted to keep his class occupied. To do so, he simply asked the class to use their slate boards and find the sum of the first 100 natural numbers. The students did what was asked of them. Namely, they began to sum the numbers $1 + 2 + 3 + 4 + 5 + 6 + \ldots$ until they reached 100. One student did not do the assignment this way and he finished immediately. That was young Carl Gauss. He decided to approach the problem in a different fashion. Rather than add the numbers in proper order:

$1 + 2 + 3 + 4 + 5 + 6 + \cdots + 98 + 99 + 100$, he decided to add the numbers in pairs:

$1 + 100 = 101$
$2 + 99 = 101$
$3 + 98 = 101$
$4 + 97 = 101$, and so on, until he reached
$48 + 53 = 101$
$49 + 52 = 101$
$50 + 51 = 101$

Of course, he did not write all this, having realized that each pair had a sum of 101, and that there were 50 such pairs. Therefore he simply multiplied $50 \times 101 = 5{,}050$ to get the sum of the numbers that Mr. Buettner requested. He was the only student in the class to even get the right answer.

When some teachers use this story as a lead-in to the lesson on finding the sum of an arithmetic series, they tend to revert to the typical textbook method for developing the formula for finding such a sum, thus defeating the fine motivational aspect of this delightful little story—one that Gauss was proud to repeat in his older days.

To use this motivational story to develop a formula for the sum of an arithmetic series, one might begin by representing the series in general terms as:

$$a + (a + d) + (a + 2d) + (a + 3d) + \cdots + (a + (n - 3)d) + (a + (n - 2)d) + (a + (n - 1)d)$$

By adding the first and the last terms and then the second and the next-to-last terms and so on, we get:

$$a + (a + (n - 1)d) = 2a + (n - 1)d$$
$$(a + d) + (a + (n - 2)d) = 2a + (n - 1)d$$
$$(a + 2d) + (a + (n - 3)d) = 2a + (n - 1)d$$

It becomes clear that there is a pattern developing—each pair yields the same sum. This is the same pattern that Gauss got when he added the numbers in pairs (as shown above). When we seek the sum of n numbers in this arithmetic series beginning with the first number a and with a common difference between terms of d, we have to sum $\frac{n}{2}$ pairs. Therefore, the sum of the series[3] is: $\frac{n}{2}(2a + (n-1)d)$.

We stress here that if one uses a story to motivate a lesson, the essence of the story should not be lost by then using another technique to develop the concept to be taught. This unfortunately happens if a teacher tells this story and then reverts to the textbook's development—if it is different from that used here—to derive the formula for the sum of an arithmetic series.

Topic: Introduction to the Pythagorean Theorem

Materials or Equipment Needed
Any medium that can present the diagram shown in figure 7.1. A chalkboard would suffice, but a computer projector would be better.

FIGURE 7.1

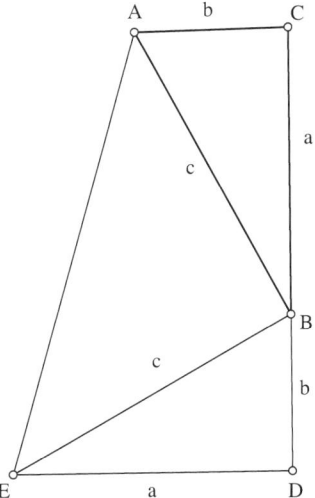

Implementation of the motivation strategy

You might begin this lesson on the Pythagorean Theorem by asking the students what Pythagoras, Euclid, and U.S. President James A. Garfield have in common. The answer is that they each proved the Pythagorean Theorem in a different way. It should be noted that two of the most famous presidents of the United States were fond of mathematics. Washington was adept at surveying and spoke favorably about mathematics, and Lincoln was known to carry a copy of Euclid's *Elements* in his saddlebag while he was still a young lawyer.

You might now tell the story of Garfield's proof and then show it. Garfield discovered the proof about five years before he became president. During this time, 1876, he was a member of Congress and hit upon the idea during a conversation about mathematics with other members of Congress. The proof was later published in the *New England Journal of Education*. After this story, with other embellishments that the teacher wishes to use here, the Garfield proof can be presented.

To begin President James A. Garfield's proof, we consider right $\triangle ABC$ with $m\angle C = 90°$; we let $AC = b$, $BC = a$, and $AB = c$. We need to show that $a^2 + b^2 = c^2$.

Select D on $\overline{BC}$ extended through point B to point D so that $BD = AC$ and we have the line segment $\overline{CDE}$. Construct $\overline{DE} \perp \overline{CBD}$ so that $DE = BC$. We can show that quadrilateral $ACDE$ is a trapezoid. Also because the triangles are congruent, Area $\triangle ABC$ = Area $\triangle BED$, and $AB = BE$.
Area trapezoid $ACDE = \frac{1}{2}CD(AC + DE) = \frac{1}{2}(a+b)\cdot(a+b) = \frac{1}{2}(a+b)^2$. Because $m\angle ABC + m\angle EBD = 90°$, $m\angle ABE = 90°$. Area $\triangle ABE = \frac{1}{2} AB \cdot BE = \frac{1}{2}c^2$. Also Area $\triangle ABC = \frac{1}{2} AC \cdot BC = \frac{1}{2}ab$.

However, Area trapezoid $ACDE$ = Area $\triangle ABE$ + 2Area $\triangle ABC$. Substituting, we get

$$\frac{1}{2}(a+b)^2 = \frac{1}{2}c^2 + 2\left(\frac{1}{2}ab\right)$$

$$(a+b)^2 = c^2 + 2ab$$

and it follows that $a^2 + b^2 = c^2$.

Topic: Introduction to the Centroid of a Triangle

Materials or Equipment Needed
A cutout map of the 48 contiguous states of the United States pasted on cardboard.

Implementation of the Motivation Strategy
The concern here is to motivate students for the lesson on the centroid of a triangle, which is the point of intersection of the three medians of the triangle. You might begin by telling them the story of how several European countries have had contests to determine the point in their country that could be considered its center. One such country is Austria, which in 1949 had such a contest, and the winner was the small resort town of Bad Aussee. They simply said that if you cut out a map of their country and mount it onto a piece of cardboard—also cut along its borders—the balancing point of the map will be their town.

The exact center of the contiguous 48 states of the United States can be similarly determined by placing a cardboard map

FIGURE 7.2

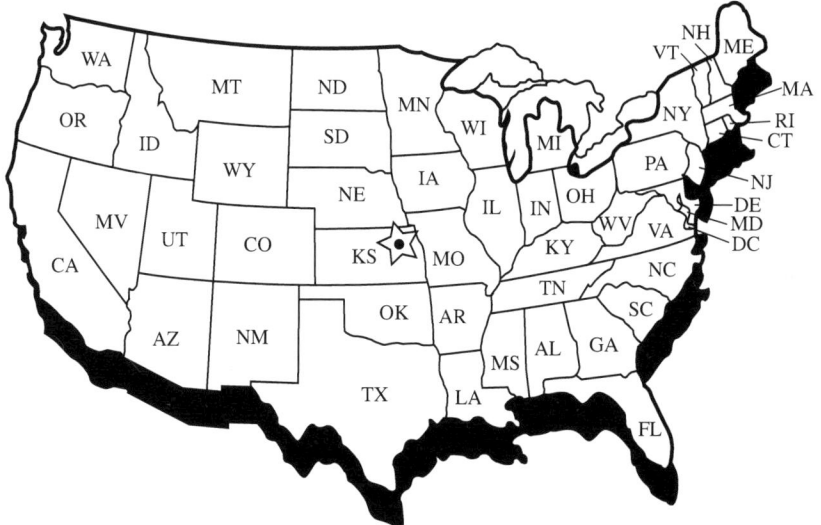

on a point to balance it. The point at which the map would be perfectly balanced is latitude 39° 50′ N, longitude 98° 35′ W, which is near the town of Lebanon in Smith County, Kansas. (See figure 7.2.)

The question for students then is how one finds the center of gravity, or centroid, of a triangle. This will lead to a discussion (and perhaps a proof) that the three medians of a triangle meet at one point, the center of gravity of the triangle. Of course, once located by construction, the centroid will be the point of balance of a cardboard cut out triangle.

Finding the center of gravity of a quadrilateral is much more complicated and is by no means as neat as finding the centroid of a triangle. This is merely a combination of finding the centroid of various triangles in the quadrilateral and then "combining" these points. Although it is not pretty to look at, we will show you how it is done so you can see the connection between the quadrilateral and triangle.

This point may be found in the following way. Let L and N be the centroids of $\triangle ABC$ and $\triangle ADC$, respectively (see figure 7.3). Let K and M be the centroids of $\triangle ABD$ and $\triangle BCD$, respectively.

FIGURE 7.3

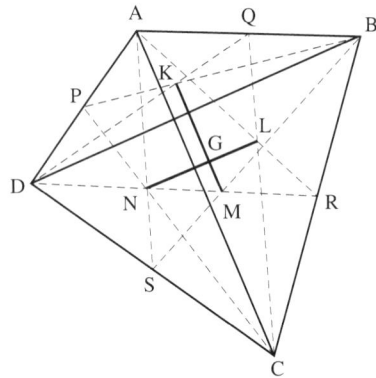

The point of intersection, G, of LN and KM is the centroid of the quadrilateral ABCD.

If you can get past the complexity of the diagram, you will see that we have simply located the centroid of the four triangles and then found the intersection of the two line segments joining them. We have then produced the analog of the triangle's centroid, or center of gravity—the point at which you could balance a cardboard quadrilateral. The centroid of a rectangle just happens to be much simpler to find. It is where you would expect it to be: at the point of intersection of the diagonals. This story and some extras should motivate students for the lesson on the concurrency of the medians of a triangle.

Topic: Introducing the Law of Sines

Materials or Equipment Needed
A worksheet as described below.

Implementation of the Motivation Strategy

Suppose you are planning a lesson on the introduction to the law of sines. You would like to develop or derive the law, and you would like to have ample time to apply the law to "practical" examples as well as the drill that typically follows the introduction of the law. So you can begin by telling a story of a

FIGURE 7.4

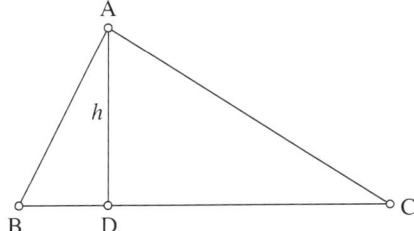

clever student, who while experimenting with the sine function in a triangle (shown in figure 7.4) with altitude h, stumbled upon a very profound trigonometric relationship. The teacher can now guide the class through the same path that this clever student found. The result is profound and the path is almost trivial!

Provide students with a worksheet with scalene triangle ABC and altitude h (figure 7.4) as follows:

$$\sin \angle ABC = \text{---}$$
$$\sin \angle ACB = \text{---}$$
$$\frac{\sin \angle ABC}{\sin \angle ACB} = \frac{}{AB}$$

Therefore,

$$\frac{}{\sin \angle ABC} = \frac{}{\sin \angle ACB}.$$

This will allow students to follow the path of the "clever student" to derive the law of sines.

Following are the expected student responses.

$$\sin \angle ABC = \frac{h}{AB}$$
$$\sin \angle ACB = \frac{h}{AC}$$
$$\frac{\sin \angle ABC}{\sin \angle ACB} = \frac{\frac{h}{AB}}{\frac{h}{AC}} = \frac{AC}{AB}$$

Therefore,

$$\frac{AC}{\sin \angle ABC} = \frac{AB}{\sin \angle ACB'}$$

which is the law of sines, and can simply be extended to the third angle of the triangle *ABC* (figure 7.4). This very concise proof will allow the teacher ample time to do a complete lesson on a topic which might otherwise require more than one lesson to introduce.

Topic: Volume and Surface Area of a Sphere

Materials or Equipment Needed
Any medium that can present the diagram shown in the figures, preferably a computer projector.

Implementation of the Motivation Strategy
Once the students have learned about the volume and surface area of the cylinder in high school, this motivator can be used to interest them in discovering the formulas for surface area and volume of a sphere. They can use these formulas to prove Archimedes' discovery.

Archimedes is considered to be one of the greatest mathematicians of all time. He lived in ancient Greece. Although he is responsible for a great many mathematical formulas and discoveries, he considered the sphere inscribed in the cylinder to be his greatest triumph.

The story is told that Archimedes was working on a diagram, which he drew in the sand with a stick. He was so engrossed in his work that when a Roman soldier told him to leave and go home, he totally ignored the soldier. After several warnings, the soldier stabbed Archimedes to death. He had just discovered that the sphere inscribed in a cylinder has a volume $\frac{2}{3}$ that of the cylinder, and the surface area of the sphere is also $\frac{2}{3}$ that of the cylinder. He considered this work to be so important and so great, that the drawing shown in figure 7.5 is inscribed on his tombstone.

FIGURE 7.5

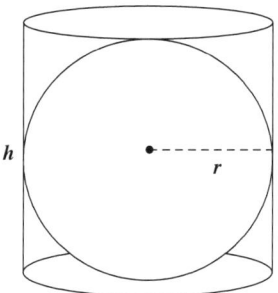

The class should find the drawing of interest. Was Archimedes correct? They already know the formula for the volume and surface area of the cylinder, and this should lead to a lesson on the volume and surface area of the sphere.

Once the students are shown (or have developed) the formulas for the volume and surface area of the sphere, they can examine the proof of Archimedes' Theorem here as follows:

Surface Area (S.A.)
The formula for the surface area of the cylinder is obtained by the formula

$$S.A. = 2\pi r^2 + 2\pi rh,$$

while the surface area of the sphere is obtained by the formula

$$S.A. = 4\pi r^2$$

Since h equals $2r$, we can rewrite the formula for the cylinder as

$$S.A. = 2\pi r^2 + 2\pi r \cdot 2r$$
$$2\pi r^2 + 4\pi r^2 = 6\pi r^2$$

The surface area of the sphere is thus $\frac{2}{3}$ that of the cylinder.

Volume
The formula for the volume of the cylinder is given by the formula

$$V = \pi r^2 h$$

The formula for the volume of the sphere is

$$V = \frac{4}{3}\pi r^3$$

Since, once again, $h = 2r$, the volume of the cylinder is

$$V = \pi r^2 \, 2r = 2\pi r^3$$

Since the volume of the sphere is $\frac{4}{3}\pi r^3$, we have the volume of the sphere is $\frac{2}{3}$ that of the cylinder.

Topic: Discovering a Prime Producing Function

Materials or Equipment Needed
A chalkboard or computer projector.

Implementation of the Motivation Strategy
The study of prime numbers and composite numbers takes place throughout middle school and high school mathematics. This motivator can be used to create a discussion on prime numbers, functions, and how to discover prime numbers.

Inform the class that many mathematicians down through the ages have searched for a function rule that will produce only prime numbers, that is, an automatic way to algebraically generate prime numbers. This is known as a "prime producing function." Although none has ever been found, you (as their teacher) will now claim that you have finally discovered one! You'll be famous! Show them the rule:

$$P = m^2 - m + 41, \text{ for positive values of } m.$$

Show how your rule produces only prime numbers:

For $m = 0$, $P = 41$, which is a prime number.
$m = 1$, $P = 41$, which is a prime number.
$m = 2$, $P = 43$, which is a prime number.
$m = 3$, $P = 47$, which is a prime number.
$m = 4$, $P = 53$, which is a prime number.
and so on.

Ask them to test your rule and be sure it yields only prime numbers.

Most of your students will laugh, and ask why it has never been discovered before. Act amazed and tell them to keep trying values for m. Students should keep substituting values for m. After a few minutes, they will probably come up with a rather puzzling result.

The function rule will produce only prime numbers for values of m from 1 through 40. However, when $m = 41$, the function yields $(41)^2 - 41 + 41$, which yields $(41)^2$, which is obviously not a prime. Thus the supposed "prime producing function" has yielded a composite and therefore is not what was originally expected. You might then mention how such inductive reasoning falls apart without a proper proof, thus leaving us still without the desired prime-producing function.

Notes

1 Alfred S. Posamentier and Noam Gordon, "An Astounding Revelation on the History of π," *Mathematics Teacher*, Vol. 77, No. 1, Jan. 1984, p. 52.
2 To read more about π, see A. S. Posamentier and I. Lehmann. π, *A Biography of the World's Most Mysterious Number* (Amherst, NY: Prometheus Books, 2004).
3 This formula is often seen as $S = \frac{n}{2}(a + l)$, where $l = a + (n - 1)d$.

8

Get Students Actively Involved in Justifying Mathematical Curiosities

In earlier chapters we showed how generating curiosity within the learner can produce effective motivation for further learning. There we emphasized some playful elements contained in mathematical puzzles or logical reasoning. In this chapter we will use mathematical curiosities to generate an interest in a particular subject. There are many number tricks being circulated through the Internet that provoke the receiver with the question: why does this happen? This type of curiosity can often be explained with simple algebra and yet can serve as a fine motivator when properly presented.

Such curiosities can be found in a number of books in the area of mathematical recreations. They should be chosen to evoke an interest in the topic of the ensuing lesson and yet be appropriate to the age group and the interests of the intended audience. Caution should be taken that these curiosities, which in and of themselves will generate interest, should not dominate the lesson; rather the justification or explanation of the curiosity should take center stage only briefly, as it introduces the topic of the day.

Topic: Introducing Probability

Materials or Equipment Needed
To do this motivational activity you will need to have a class of at least 30 students. Even better would be to do this motivational activity in a school where there are ten classes of about 30 students each. If this is not possible, then gather ten lists of 30 people each, along with their birth dates (excluding the year).

Implementation of the Motivation Strategy
Begin by asking the class what they think the chances (or probability) are of two classmates having the same birth date (month and day only) in their class of about 30+ students. Students usually begin to think about the likelihood of 2 people having the same date out of a selection of 365 days (assuming no leap year). Perhaps 2 out of 365? Ask them to consider the "randomly" selected group of the first 30 presidents of the United States. This can be one of the lists of 30 people (with birth dates) you have prepared before class. They may be astonished that two have the same birth date: James K. Polk (November 2, 1795) and Warren G. Harding (November 2, 1865). The class will probably be surprised to learn that for a group of 30, the probability that two members will have the same birth date is greater than 70 percent. Students may wish to try their own experiment by visiting 10 nearby classrooms to check on date matches. For groups of 30, the probability is that there will be a match of birth dates in 7 of these 10 rooms. How does this incredible and counterintuitive probability evolve? This is now highly motivating for the class to delve into this lesson on probability. To guide students to justify these rather curious and unexpected probabilities, consider the following:

What is the probability that one student matches his own birth date? Clearly, certainty, or 1.

This can be written as $\frac{365}{365}$.

The probability that another student does *not* match the first student is $\frac{365-1}{365} = \frac{364}{365}$.

The probability that another student does *not* match the first and second students is $\frac{365-2}{365} = \frac{363}{365}$.

The probability of all 30 students *not* having the same birth date is the product of these probabilities:

$$p = \frac{365}{365} \cdot \frac{365-1}{365} \cdot \frac{365-1}{365} \cdot \ldots \cdot \frac{365-29}{365}.$$

Since the probability (q) that two students in the group have the same birth date and the probability (p) that two students in the group do not have the same birth date is a certainty, the sum of those probabilities must be 1. Thus, $p + q = 1$.

In this case, $q = 1 - \frac{365}{365} \cdot \frac{365-1}{365} \cdot \frac{365-2}{365} \cdot \ldots \cdot \frac{365-28}{365} \cdot \frac{365-29}{365} \approx 0.7063162427192686$. In other words, the probability that there will be a birth date match in a randomly selected group of 30 people is somewhat greater than $\frac{7}{10}$. This is quite unexpected when one considers there were 365 dates from which to choose. Students may wish to investigate the nature of the probability function. Here are a few values to serve as a guide:

Number of people in group	Probability of a birth date match	Number of people in group	Probability of a birth date match
10	0.1169481777110776	45	0.9409758994657749
15	0.2529013197636863	50	0.9703735795779884
20	0.4114383835805799	55	0.9862622888164461
25	0.5686997039694639	60	0.994122660865348
30	0.7063162427192686	65	0.9976831073124921
35	0.8143832388747152	70	0.9991595759651571
40	0.891231809817949		

Students should notice how quickly almost-certainty is reached.

Were one to do this with the death dates of the first 30 presidents, one would notice that two died on March 8 (Millard Fillmore and William H. Taft) and three presidents died on the fourth of July (Adams, Jefferson, and Monroe). Above all, this motivating start to the topic of probability should serve as an eye-opener about relying on intuition too much.

Topic: A Lesson on Digit Problems and Place Value

Materials or Equipment Needed
Bring to class enough previously used restaurant receipts so that each group in the class can have one of them.

Implementation of the Motivation Strategy
Present your class with the following situation. It is often very useful to be able to look at a number and quickly determine if it is divisible by another number. For example, you go to a restaurant with two friends, and when the check comes you decide to add a tip and then divide the bill equally in thirds so that each person pays the same amount. How can we determine divisibility by 3 without actually dividing?

Have each of the groups first calculate an 18 percent tip on the total bill. Then ask them to modify the tip as little as possible so that the total is divisible by 3. This should probably be done without actually doing the division. The ensuing lesson on divisibility by 3 will answer this challenge.

Now we present a useful rule that states that if the sum of the digits of a number is divisible by 3, then the number is also divisible by 3. For example, if we want to determine if the number 537 is divisible by 3, using this rule, we simply find the sum of the digits of this number: $5 + 3 + 7 = 15$, and check the sum. Since 15 is divisible by 3, the number 537 is also divisible by 3.

We must be able to justify this rule in order to accept it. Students should briefly attempt to find some form of justification. After a short while students will be eager to be guided to this justification—that is, motivated to justify a place-value issue.

Have the students represent a general 3-digit number, say *htu*, in place-value form: $100h + 10t + u$. Then ask them how the number 3 could be worked into this expression. A clever student might come up with the creative response to change this expression to: $(99 + 1)h + (9 + 1)t + u$. This can then be rewritten as: $99h + 9t + h + t + u$. Since $99h + 9t = 9(11h + t)$ is always divisible by 3, students should then see that the entire expression, $99h + 9t + h + t + u$ would be divisible by 3, when the remaining part, $h + t + u$ is divisible by 3. In other words, when the sum of the digits is divisible by 3, the number is also divisible by 3.

It is easy to see that the same rule will also hold for divisibility by 9. In more general terms, the rule holds for one less than the base and its factors.

From here the teacher may lead the class to consider analogous divisibilities in other bases as a beginning to considering other base representations. A teacher may want to take this further to other problems involving the expanded form of base-10 numbers.

Topic: Application of Digit Problems in Algebra

Materials or Equipment Needed
Nothing special is required for this motivational activity, except a chalkboard or any other method of writing something for the entire class to see.

Implementation of the Motivation Strategy
Begin the class in the unusual way of asking every student to select a 3-digit number, where the unit and hundreds digits are not the same. This mind-boggling activity involves number properties that are exceptional, leaving students with a strong desire to discover why this is so—hence highly motivated students.

We provide you with a simulated version of what is to be done with the class. In bold print are the instructions for the students to follow.
Choose any 3-digit number (where the unit and hundreds digits are not the same).

We will do it with you here by arbitrarily selecting **825.**

Reverse the digits of the number you have selected.

We continue here by reversing the digits of 825 to get **528**.

Subtract the two numbers (naturally, the larger minus the smaller).

Our calculated difference is **825 − 528 = 297**.

Once again, reverse the digits of this difference.

Reversing the digits of 297 we get the number **792**.

Now, add your last two numbers.

We then add the last two numbers to get: **297 + 792 = 1089**.

Each of you should have the same result, even though your starting number was different from your classmates'.

Students will probably be astonished that regardless of which number was selected at the beginning, the same result was reached by all, namely, 1089.

How does this happen? Is this a "freak property" of this number? Did we do something illegitimate in our calculations?

Did we assume that any number we chose would lead us to 1089? How could we be sure? Well, we could try all possible 3-digit numbers to see if they work. That would be tedious and not particularly elegant. This is where you can embark on the lesson—applications of algebra with digit problems.

Following is a possible method to serve as a guide for the teacher.

We shall represent the arbitrarily selected 3-digit number, htu, as $100h + 10t + u$, where h represents the hundreds digit, t represents the tens digit, and u represents the units digit.

Let $h > u$, which would be the case either in the number you selected or the reverse of it.

In the subtraction, $u - h < 0$; therefore, take 1 from the tens place (of the minuend), making the units place $10 + u$.

Since the tens digits of the two numbers to be subtracted are equal, and 1 was taken from the tens digit of the minuend, then the value of this digit is $10(t-1)$. The hundreds digit of the minuend is $h-1$, because 1 was taken away to enable subtraction in the tens place, making the value of the tens digit $10(t-1) + 100 = 10t(t+9)$.

We can now do the first subtraction:

$$
\begin{array}{lll}
100(h-1) & +10(t+9) & +(u+10) \\
100u & +10t & +h \\
\hline
100(h-u-1) & +10(9) & +u-h+10
\end{array}
$$

Reversing the digits of this difference gives us:

$$100(u - h + 10) + 10(9) + (h - u - 1)$$

Now adding these last two expressions gives us:

$$100(9) + 10(18) + (10 - 1) = \underline{1089}.$$

The algebra enables us to inspect the arithmetic process, regardless of the number.

There is a particular beauty in the number **1089**. As a little extra, you may want to entertain the class with another oddity of this now famous number. Let's look at the first ten multiples of 1089.

$$
\begin{array}{l}
1089 \cdot 1 = 1089 \\
1089 \cdot 2 = 2178 \\
1089 \cdot 3 = 3267 \\
1089 \cdot 4 = 4356 \\
1089 \cdot 5 = 5445 \\
1089 \cdot 6 = 6534 \\
1089 \cdot 7 = 7623 \\
1089 \cdot 8 = 8712 \\
1089 \cdot 9 = 9801
\end{array}
$$

Do you notice a pattern among the products? Look at the first and ninth products. They are reverses of one another. The

126 ◆ Get Students Actively Involved in Justifying Mathematical Curiosities

second and the eighth are also reverses of one another. And so the pattern continues, until the fifth product is the reverse of itself, known as a palindromic number.

Topic: Introducing the Base-2 Number System

Materials or Equipment Needed
Teacher will need to have a large version of the chart shown in figure 8.1, or individual sheets with the chart for each student in the class.

Implementation of the Motivation Strategy
In middle school, students are introduced to number bases other than base 10. One way to get them interested in number bases is to use this activity, where the teacher will present a "baffling" puzzle whose results students will be eager to justify. This activity should motivate your students to become interested in number bases other than base 10.

Prepare a set of "Mind-Reading Cards," as shown in figure 8.1. Ask a student to select any number from 1 through 63 and write it on a slip of paper. Fold the paper and give it to another student to hold without looking at it yourself.

Ask the student to examine the Mind-Reading Cards one at a time, and tell whether or not his or her number appears on each card in turn. For the teacher to guess the secret number, all that needs to be done is add the first number on each card on which the student's number appears.

You should then be able to tell the student which number was picked. Check it against the slip of paper on which the student

FIGURE 8.1

1 3 5 7	2 3 6 7	4 5 6 7	8 9 10 11	16 17 18 19	32 33 34 35
9 11 13 15	10 11 14 15	12 13 14 15	12 13 14 15	20 21 22 23	36 37 38 39
17 19 21 23	18 19 22 23	20 21 22 23	24 25 26 27	24 25 26 27	40 41 42 43
25 27 29 31	26 27 30 31	28 29 30 31	28 29 30 31	28 29 30 31	44 45 46 47
33 35 37 39	34 35 38 39	36 37 38 39	40 41 42 43	48 49 50 51	48 49 50 51
41 43 45 47	42 43 46 47	44 45 46 47	44 45 46 47	52 53 54 55	52 53 54 55
49 51 53 55	50 51 54 55	52 53 54 55	56 57 58 59	56 57 58 59	56 57 58 59
57 59 61 63	58 59 62 63	60 61 62 63	60 61 62 63	60 61 62 63	60 61 62 63

had written his or her number. Students will want to know *how* this little curiosity was accomplished. This leads into a lesson on expressing numbers in bases other than 10, in this case, base 2.

In base 2, the only symbols we can use are 1 and 0. Just as in our system (base 10), the *position* of the 1s and 0s determines their value. The first number on each card, in order, is 1, 2, 4, 8, 16, and 32. This represents $2^0 = 1$, $2^1 = 2$, $2^2 = 4$, and so on through $2^5 = 32$.

A number appearing on a card is the same as placing a "1" in the appropriate position. The number not appearing on the card is equivalent to placing a "0" in that position. In this manner, the number 23 is written in base 2 as **10111** and appears on the 4 cards each representing a different position of the place value in base 2. The number is $\mathbf{1}(16) + \mathbf{0}(8) + \mathbf{1}(4) + \mathbf{1}(2) + \mathbf{1}(1)$. Thus, 23 will appear on the cards headed by 16, 4, 2, and 1. Adding these, we get $16 + 4 + 2 + 1 = 23$.

Topic: Application of Digit Problems in Algebra, or Using Algebra to Justify an Arithmetic Peculiarity

Materials or Equipment Needed
A chalkboard or any other medium for students to follow the development.

Implementation of the Motivation Strategy
This motivational device will begin with the students being shown an arithmetic curiosity then trying to find other examples of it and finally attempting to explain its underlying "secret."

Tell students that there are certain number pairs that yield the same product even when both numbers have their digits reversed. For example, $12 \times 42 = 504$ and, when we reverse the digits of each of the two numbers, we get $21 \times 24 = 504$. The same is true for the number pair 36 and 84, since $36 \times 84 = 3024 = 63 \times 48$.

At this point students may wonder if this will happen with any pair of numbers. The answer is that it will only work with 14 pairs of numbers:

$$12 \times 42 = 21 \times 24 = 504$$
$$12 \times 63 = 21 \times 36 = 756$$

$12 \times 84 = 21 \times 48 = 1008$
$13 \times 62 = 31 \times 26 = 806$
$13 \times 93 = 31 \times 39 = 1209$
$14 \times 82 = 41 \times 28 = 1148$
$23 \times 64 = 32 \times 46 = 1472$
$23 \times 96 = 32 \times 69 = 2208$
$24 \times 63 = 42 \times 36 = 1512$
$24 \times 84 = 42 \times 48 = 2016$
$26 \times 93 = 62 \times 39 = 2418$
$34 \times 86 = 43 \times 68 = 2924$
$36 \times 84 = 63 \times 48 = 3024$
$46 \times 96 = 64 \times 69 = 4416$

A careful inspection of these 14 pairs of numbers will reveal that in each case the product of the tens digits of each pair of numbers is equal to the product of the units digits. You can now satisfy students with an algebraic examination of the peculiarity. We have as follows for the numbers z_1, z_2, z_3, and z_4:

$$z_1 \times z_2 = (10a + b) \times (10c + d) = 100ac + 10ad + 10bc + bd$$

and

$$z_3 \times z_4 = (10b + a) \times (10d + c) = 100bd + 10bc + 10ad + ac,$$

where a, b, c, d represent any of the ten digits: 0, 1, 2, . . ., 9, where $a \neq 0$ and $c \neq 0$.
We must have $z_1 \times z_2 = z_3 \times z_4$

$100ac + 10ad + 10bc + bd = 100bd + 10bc + 10ad + ac$

$100ac + bd = 100bd + ac$

$99ac = 99bd$

$ac = bd$, which we observed earlier.

Topic: Applying the Trigonometric Angle Sum Function

Materials or Equipment Needed
A dynamic geometry program would be useful—such as Geometer's Sketchpad or GeoGebra.

Implementation of the Motivation Strategy

Begin the lesson with the statement (which might surprise students) that three right triangles can be found so that if you calculate the size of the smallest angle of each triangle and add the three measurements, the sum is 90°. To justify this curiosity should motivate students. They should be guided to focus on an angle sum formula from trigonometry.

Begin by using Geometer's Sketchpad or GeoGebra to draw three right triangles with sides of the following lengths (figure 8.2):

$$3, 4, 5; \quad 8, 15, 17; \quad 36, 77, 85.$$

In each case, the size of the triangle is not important, just the shape. That is, the relationship among the sides of each triangle must be as indicated above.

The task at hand is to find the sum of the measures of the smallest angles of each of these right triangles. Since the smallest angle of a triangle is opposite the smallest side, we seek to find the sum of $m \angle C$, $m \angle F$, and $m \angle J$.

From $\triangle ABC$ we get the trigonometric ratios: $\sin C = \frac{3}{5}$, $\cos C = \frac{4}{5}$. Similarly, from $\triangle DEF$ $\sin F = \frac{8}{17}$, and $\cos F = \frac{15}{17}$. Now, $\sin(C+F) = \sin C \cdot \cos F + \cos C \cdot \sin F = \frac{3}{5} \cdot \frac{15}{17} + \frac{4}{5} \cdot \frac{8}{17} = \frac{77}{85}$.

FIGURE 8.2

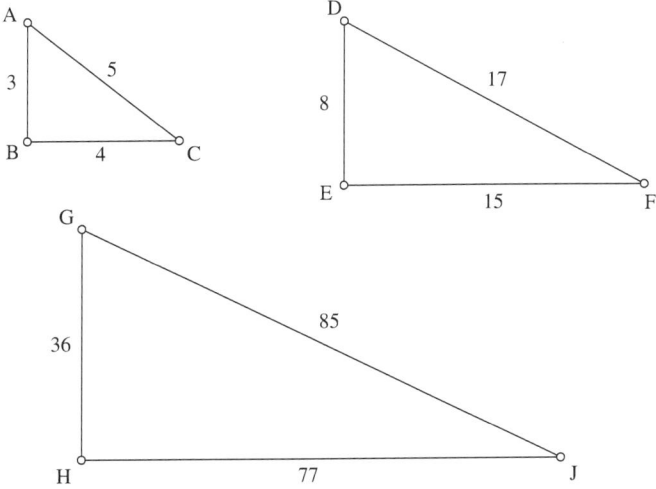

We notice from the third triangle that $\cos J = \frac{77}{85} = \sin(90 - J)$.

Since both are equal to $\frac{77}{85}$, $\sin(C + F) = \sin(90 - J)$. We conclude that $C + F = 90 - J$, or $C + F + J = 90$, which was to have been found.

One might try to "verify" this result with the Geometer's Sketchpad. Here are the measures of the angles:

$$m\angle ACB \approx 37°$$
$$m\angle DFE \approx 28°$$
$$m\angle GJH \approx 25°$$

It should be emphasized that the measures are only approximate, whereas the relationship $m\angle ACB + m\angle DFE + m\angle GJH = 90°$ is exact.

This problem points out a nice use of the angle sum function.

Now try to find another triple of such right triangles.[1] Are there an infinite number of such triples of right triangles?

Note

1 Here is another set of three right triangles where this will hold true: (3, 4, 5), (5, 12, 13), and (33, 56, 65).

9

Employ Teacher-Made or Commercially Prepared Materials

We know that many of our students learn best from an actual "hands-on" approach. They need a physical object in their hands to begin to comprehend what is taking place. Motivation can be achieved by presenting the class with a hands-on approach, using concrete materials of an unusual nature. This may include teacher-made materials, such as models of geometric shapes, geo strips, specifically prepared PowerPoint slides, or practical devices that illustrate a specific geometric principle. Some excellent commercially prepared materials are available, ranging from geometric models to videos of various kinds. Many computer programs exist that lead to fruitful discussions in class. Materials selected should be reviewed thoroughly and their presentation carefully planned so as to motivate students for the lesson and not to detract attention from it. Almost any sort of "prop" can be used to engage student interest.

Unusual materials rouse student interest and can thus heighten the desire to analyze the subject further. Motivation psychology calls this underlying behavior the "curiosity motive" and suggests this is innate in humans. Modern developmental psychology considers curiosity an important driving force for children and adolescents in their relationship with their environment.

It is fascinating to observe how students who normally show reserve and little initiative suddenly become interested when the teacher produces materials of an unusual nature. The impetus that accompanies this first surprise often continues beyond the first lesson and throughout the whole teaching sequence. Time and again students report at home about their unusual school experiences and what surprising things they did in their math class. This has the advantage that they review their learning and what they have seen and heard at school. Furthermore, this helps to show that math is by no means the dry subject it is so often considered; on the contrary, it has surprising and unusual aspects which are there for us to discover.

Students can become passionately inspired through the clever use of well-chosen materials. Teachers and their skills are challenged here to be creative and find relevant items to bring to class. Finally, as with all other methods, teachers must take into consideration the context, time, and frequency of implementation. Finding the appropriate balance between too much and too little is not always easy, but it will ultimately determine the success of further lessons.

Topic: Introducing the Concept of a Function

Materials or Equipment Needed
The usual medium to present a problem—a computer projector or chalkboard would suffice; however a toy archery set might be helpful for demonstration purposes.

Implementation of the Motivation Strategy
If the teacher brings to the class a toy archery set to dramatize the motivational activity presented here, just the sight of the unusual items will arouse—or motivate—students to discover what the ensuing lesson is going to be about. The following is actually a strong introduction to the concept of functions that has been motivated by the archery set and by the discussion that follows.

In mathematics, finding concrete analogues to represent abstract concepts is not always easy. One example where a physical model can be used to explain an abstract concept is in the development of the notion of a function.

We will use the model of a bow shooting arrows at a target. The arrows will represent the *domain* and the target represents the *range*. The bow (and its aiming) is the *function*. Since an arrow can be used only once,[1] we know that the elements in the domain can be used only once. The bow can hit the same point on the target more than once. Therefore, points in the range can be used more than once. This is the definition of a function: a *mapping* of all elements of one set onto another, with the elements of the first set used exactly once. Some points on the target may never be hit by an arrow, yet all the arrows must be used. Analogously, some elements in the range may not be used, but all elements in the domain must be used. Or conversely, through a mapping (or a "pairing") of all elements in the domain, some elements in the range may not be used.

When all points on the target (the range) are hit,[2] then the function (or mapping) is called an *onto function*.

When each point on the target is used only once, then the function is called a *one-to-one function*.

When each point on the target is used exactly once (once and only once) then the function is called a *one-to-one onto function*, or may be called a *one-to-one correspondence*.

Using the bow-shooting-arrows-to-a-target analogy to represent the concept of a function enables the learner to conceptualize this abstract notion in a way that should instill permanent understanding. Using a toy dart set is preferred for classroom use rather than the archery set!

Topic: Developing the Formula for the Area of a Circle

Materials or Equipment Needed
Construction paper prepared as shown in figures 9.1 and 9.2.

Implementation of the Motivation Strategy
Students are often "told" that the area of a circle is found by the formula $A = \pi r^2$. Too often, they are not given an opportunity to discover where this formula may have come from or how it

relates to other concepts they have learned. It is not only entertaining, but also instructionally sound, to have the formula evolve from previously learned concepts. Assuming that the students are aware of the formula for finding the area of a parallelogram, this motivator presents a nice justification for the formula for the area of a circle.

This motivational activity will use the teacher-prepared materials—a convenient size circle drawn on the piece of cardboard or construction paper, divided into 16 equal sectors (see figure 9.1). This may be done by marking off consecutive arcs of 22.5° or by consecutively dividing the circle into two parts, then four parts, then bisecting each of these quarter arcs, and so on. These sectors, shown in figure 9.1, are then to be cut apart and reassembled in the manner shown in figure 9.2.

This placement suggests that we have a figure that approximates a parallelogram. That is, were the circle cut into more sectors, then the figure would approach a true parallelogram. Let us assume it is a parallelogram. In this case the base would have length $\frac{1}{2}C$, where $C = 2\pi r$ (r is the radius). The area of

FIGURE 9.1

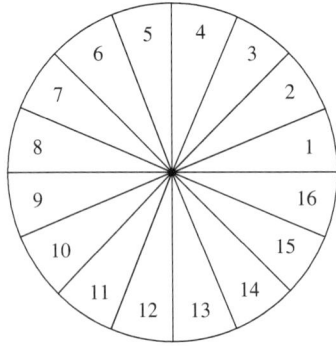

FIGURE 9.2

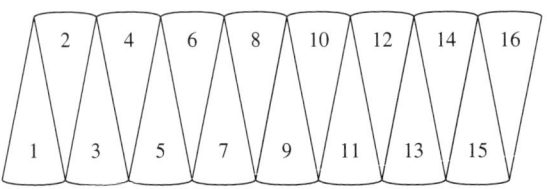

the parallelogram is equal to the product of its base and altitude (which here is r). Therefore, the area of the parallelogram $= \left(\frac{1}{2}c\right)r = \frac{1}{2}(2\pi r)(r) = \pi r^2$, which is the commonly known formula for the area of a circle. This should certainly impress your students to the point where this area formula begins to have some intuitive meaning. This motivational activity will have a profound impression on students who have seen the circle area formula many times and need refresher motivation.

Topic: Developing the Sum of the Angles of a Triangle

Materials or Equipment Needed
Any paper cut into a triangular shape.

Implementation of the Motivation Strategy
In middle school, the students have worked with triangles that are equilateral, isosceles, right, acute, obtuse, and so on. What do they all have in common? Besides having three sides, every triangle has angle measures that total 180 degrees. This motivator provides simple teacher-made materials to act as an incentive for the students to explore this relationship and leads to a lesson designed to prove and apply the theorem.

Give each group of students, or individual student, a triangle cut from cardboard or paper. Have the students locate the midpoints of two sides. In figure 9.3, points D and E are the midpoints of sides AB and AC. Then folding the top vertex of the triangle

FIGURE 9.3

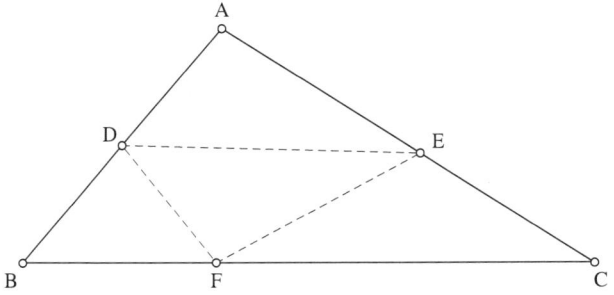

along *DE* will allow the vertex *A* to land on *BC* as shown in figure 9.4. Next fold vertex *B* over to meet point *A* (figure 9.5) and similarly on the other side having vertex *C* also meet at point *A* (figure 9.6). Therefore, the sum of the angles of a triangle is 180°.

Another option would be to tear off the three angles of the original triangle and place them together to show a straight line sum. However, there are interesting relationships that can be shown with paper-folding.

To be sure that students see the general nature of this demonstration, some groups should have an equilateral triangle, some an isosceles triangle, some a right triangle, and others a scalene triangle so that the result can be generalized.

FIGURE 9.4

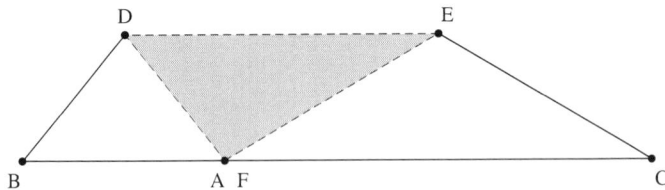

FIGURE 9.5

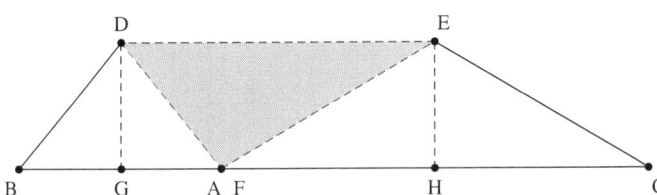

FIGURE 9.6

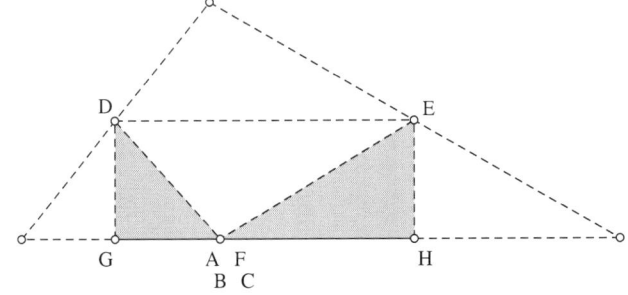

Topic: Introducing the Triangle Inequality

Materials or Equipment Needed
A box of spaghetti and a chalkboard.

Implementation of the Motivation Strategy
Walking into the classroom with a box of spaghetti will arouse interest or motivation among the students and in this case will solidify the concept of the triangle inequality. Carefully give each student ten spaghetti sticks. Tell students to take each spaghetti stick and break it into three parts. Then they are to place the three pieces on their desk and try to form a triangle. They are to keep a log to note if they were able to form a triangle for each attempt with a spaghetti stick. Reviewing their ten attempts should lead students to the conclusion that only when the sum of the lengths of any two sides is greater than the length of the third side will it be possible to construct a triangle.

Students have probably heard that "the shortest distance between two points is a straight line." We can use this fact to arrive at the triangle inequality:

The sum of the lengths of any two sides of a triangle must be greater than the length of the third side.

In figure 9.7, the shortest distance between points A and B is segment AB, that is, $AC + CB > AB$.

The proof of this relationship (or theorem) can be done rather simply.

Consider the triangle ABC (figure 9.8), and choose the point D on ray CA so that $AD = AB$.

FIGURE 9.7

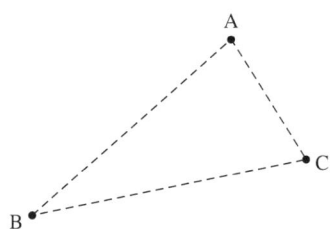

FIGURE 9.8

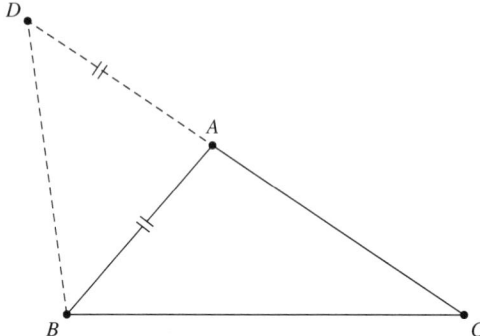

Since in isosceles triangle DAB, $\angle ADB \cong \angle ABD$, $m\angle DBC > m\angle ADB$. It then follows that (for $\triangle DBC$), $DC > BC$, because in a triangle the greater side is opposite the greater angle. However, since $AD = AB$, $DC = AC + AB$. Therefore, $AC + AB > BC$, which is what was to be proved.

Topic: Extending the Pythagorean Theorem

Materials or Equipment Needed
Construction paper prepared as shown below.

Implementation of the Motivation Strategy
Begin by informing the students that the Pythagorean Theorem has been celebrated by well over 520 different proofs,[3] some of which were done by Pythagoras (ca. 570 BC–510 BC), Euclid (ca. 365 BC–ca. 300 BC), Leonardo da Vinci (1452–1519), Albert Einstein (1879–1955), and U.S. President James A. Garfield (1831–1881)—when he was a member of the House of Representatives. After you mention that the sum of the areas of the squares on the legs of right triangle is the same as the area of the square on the hypotenuse—and perhaps prove it in any of several ways—you might motivate the class by showing them that this is not limited to "squares" and can also be said for the areas of any similar polygons on the sides of the right

FIGURE 9.9   **FIGURE 9.10**

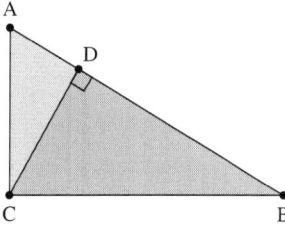 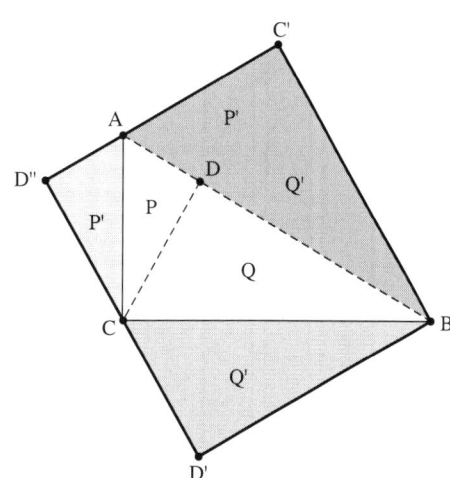

triangle. We shall use similar right triangles, shown in figure 9.9, and place them on the sides of a right triangle (figure 9.10), and then "visually" prove the Pythagorean Theorem in a manner that could be seen as intuitive.

Consider the right $\triangle ABC$ with right angle ACB and altitude CD. In figure 9.9, we cover each of the three similar right triangles, $\triangle ADC$, $\triangle CDB$, and $\triangle ACB$, with a *congruent* triangle. We can plainly see that Area $\triangle ADC$ + Area $\triangle CDB$ = Area $\triangle ACB$.

We then flip them over on their respective hypotenuses as is shown in figure 9.10, so that we can conclude that the sum of the areas of the similar right triangles ($\triangle AD''C$, and $\triangle CD'B$) on the legs of the right triangle $\triangle ABC$ is equal to the area of the similar right triangle ($\triangle AC'B$) on the hypotenuse. As we said earlier, since the ratio of the areas of any similar polygons is equal to the ratio of the squares of the corresponding sides, we get $(AC)^2 + (CB)^2 = (AB)^2$, which is the Pythagorean Theorem—this proves it!

This teacher-made model will go a long way to motivate students to seek out other nontraditional methods to prove the Pythagorean Theorem.

Topic: Introducing the Pythagorean Theorem

Materials or Equipment Needed
Cardboard cut-outs as shown in figures 9.11 and 9.12, and a rope with 12 equally spaced knots as shown in figure 9.15.

Implementation of the Motivation Strategy
As soon as the teacher shows some unusual materials at the start of class, students typically get curious as to what will be coming in the ensuing lesson. This motivation is one of many types that can be used to introduce the Pythagorean Theorem. This coupled with a bit of light history of the theorem would be further motivating for the students.

We now focus on the geometric relationship that made Pythagoras famous in today's world and which, of course, bears his name. We would do well to consider his prominent role (or that of his society) in the development of this amazing relationship. Although the relationship was already known before Pythagoras, it is appropriate that the theorem should be named for him, since Pythagoras (or one of the Pythagoreans) was the first to give a proof of the theorem—at least as far as we know. Historians suppose that he used the squares as shown in figures 9.11 and 9.12—perhaps inspired by the pattern of floor tiles. We will briefly demonstrate the proof here.

FIGURE 9.11   **FIGURE 9.12**

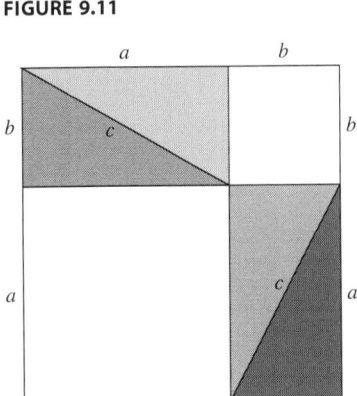 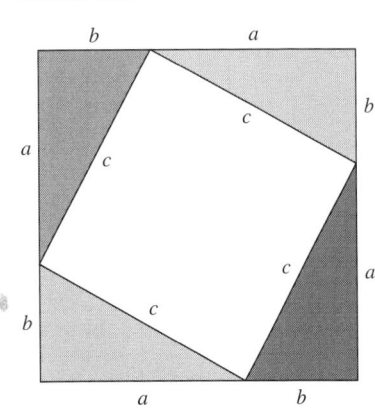

To show that $a^2 + b^2 = c^2$, you need only subtract the four right triangles, with sides a, b, and c, from each of the two larger squares, so that in figure 9.11 you end up with two squares (figure 9.13) whose area sum is $a^2 + b^2$, and in figure 9.12 you end up with a square of area c^2 (figure 9.14). Therefore, since the two original squares were the same size and we subtracted equal quantities from each, we can conclude that $a^2 + b^2 = c^2$, which is shown in figures 9.13 and 9.14 with the two figures having the same area.

To prove a theorem is one thing, but to come up with the idea establishing this geometric relationship is quite another. It is likely that Pythagoras learned about this relationship on his study trip to Egypt and Mesopotamia, where this concept was known and used in construction for special cases.

During his travels to Egypt, Pythagoras probably witnessed the measuring method of the so-called Harpedonapts (rope stretchers). They used ropes tied with 12 equidistant knots to create a triangle with two sides of length 3 and 4 units and a third side of 5 units, knowing that this enabled them to "construct" a right angle. (See figure 9.15.)

FIGURE 9.13

FIGURE 9.14

FIGURE 9.15

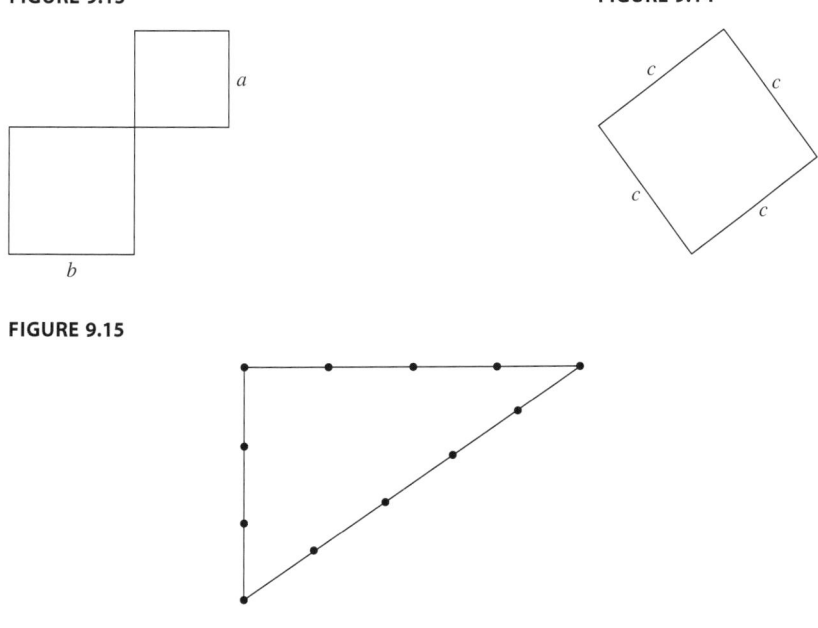

They applied this knowledge to survey the banks of the river Nile after the annual floods in order to rebuild rectangular fields for the farmers. They also employed this method in laying the foundation stones of temples. To the best of our knowledge, the Egyptians did not know of the generalized relationship given to us by the Pythagorean Theorem. They seem to have only known about the special case of the triangle with side-lengths 3, 4, and 5, which produced a right triangle. This was arrived at by experience, and not by some sort of formal proof.

This bit of historical background coupled with the unusual teacher-made materials should serve as one of many fine motivating introductions to the Pythagorean Theorem.

Topic: Introduction to Angle Measurement with a Circle by Moving the Circle

Materials or Equipment Needed

Any medium that can present the diagram shown in figure 9.16; (suggested) a cardboard with two dark strings fastened as shown and a cut-out cardboard circle with an angle marked on it equal to the angle of the two strings fastened to the large cardboard.

Implementation of the Motivation Strategy

The presentation will be quite motivating since most of it is with physical models—either cardboard or with the computer. This lesson is designed to prove the theorems on measuring angles related to a circle using a physical model, such as cardboard, string, and a pair of scissors, or using the Geometer's Sketchpad software program. Rather than following the usual textbook method of treating each of the theorems as separate entities, this lesson will take care of all the theorems with one procedure. The applications that follow this lesson may be handled either together (as we are presenting the theorems here) or individually over a few lessons following this one.

We will demonstrate that all of the measurements of the various angles related to a circle can be carried out very nicely by cutting out a circle from a piece of cardboard and drawing a convenient inscribed angle on it (figure 9.16). The measure of that angle should be the same as that formed by two pieces of string, which are affixed to a rectangular piece of cardboard as in figure 9.16.

FIGURE 9.16

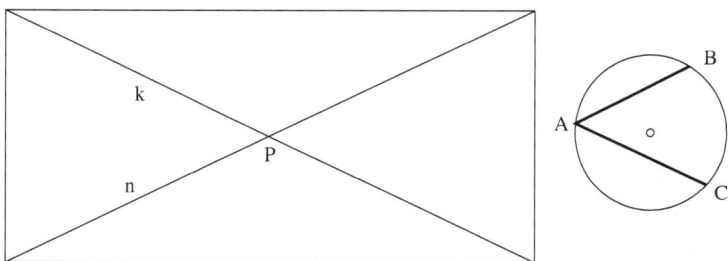

The lesson prior to this one should have proved the theorem establishing that the measure of an inscribed angle of a circle is one-half the measure of the intercepted arc.

By moving the circle to various positions, we will be able to find the measure of an angle formed by:

- Two chords intersecting inside the circle (but not at its center)
- Two secants intersecting outside the circle
- Two tangents intersecting outside the circle
- A secant and a tangent intersecting outside the circle
- A chord and a tangent intersecting on the circle

We begin with demonstrating the relationship between the arcs of the circle and the angle formed **by two chords intersecting inside the circle** (but not at its center). Place the cardboard circle into a position so that $\overline{AB} \| n$ and $\overline{AC}$ are on k, as in figure 9.17.

Notice that $m\angle A = \frac{1}{2} m\widehat{BEC}$, and $m\angle A = m\angle EPC$. Therefore $m\angle P = \frac{1}{2} m\widehat{BEC} = \frac{1}{2}\left(m\widehat{BE} + m\widehat{EC}\right)$. But, since parallel lines cut off congruent arcs on a given circle, $m\widehat{BE} = m\widehat{AF}$. It then follows that $m\angle P = \frac{1}{2}\left(m\widehat{AE} + m\widehat{EC}\right)$, which shows the relationship of the angle formed by two chords, $\angle P$, and its intercepted arcs, $\widehat{AF}$ and $\widehat{EC}$.

FIGURE 9.17

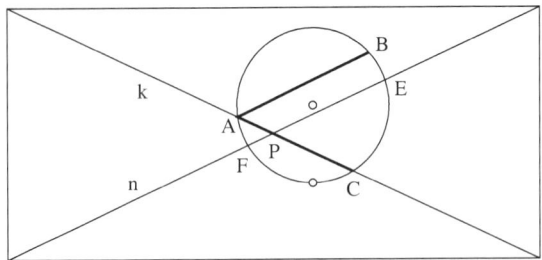

FIGURE 9.18

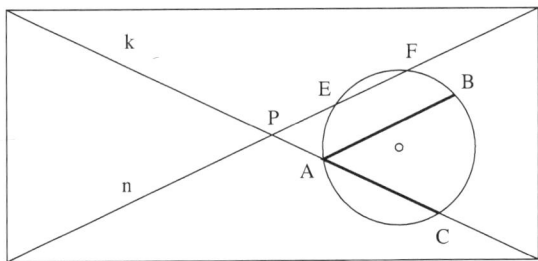

Consider next the angle formed by **two secants intersecting outside the circle.** Place the cardboard circle into the position shown in figure 9.18.

Begin by remembering that $m\angle P = \tfrac{1}{2}m\widehat{BC}$ and $m\angle FPC = m\angle A$. Since $m\widehat{AE} = m\widehat{BF}$, we can add and subtract it to the same quantity without changing the value of the original quantity. Thus, $m\angle P = \tfrac{1}{2}\left(m\widehat{BC} + m\widehat{BF} - m\widehat{AE}\right) = \tfrac{1}{2}\left(m\widehat{FBC} - m\widehat{AE}\right)$.

In a similar way we can demonstrate the relationship between an angle formed by **two tangents intersecting outside the circle** and its intercepted arcs. We move the cardboard circle into the position shown in figure 9.19.

In this case the equality of arcs $\widehat{AE}$ and $\widehat{BE}$ as well as that of arcs $\widehat{AF}$ and $\widehat{CF}$ is key to demonstrating the desired relationship.

We have $m\angle P = m\angle A = \tfrac{1}{2}m\widehat{BC} = \tfrac{1}{2}\left(m\widehat{BE} + m\widehat{BC} + m\widehat{CF} - m\widehat{AE} - \widehat{AF}\right) = \tfrac{1}{2}\left(m\widehat{EBCF} - m\widehat{EAF}\right)$.

Again, by sliding the cardboard circle to the following position (see figure 9.20) we can find the measure of the angle formed **by a tangent and a secant intersecting outside the circle.**

This time we rely on the equality of arcs $\widehat{AE}$ and $\widehat{BE}$. We get the following by adding and subtracting these equal arcs:

$$m\angle P = m\angle A = \frac{1}{2}m\widehat{BC} = \frac{1}{2}\left(m\widehat{BC} + m\widehat{BE} - m\widehat{AE}\right) = \frac{1}{2}\left(m\widehat{EBC} - m\widehat{AE}\right).$$

To complete the various possibilities of positions for the cardboard circle, place it so that we can find the relationship between **an angle formed by a chord and a tangent intersecting at the point of tangency** and its intercepted arc. See figure 9.21.

FIGURE 9.19

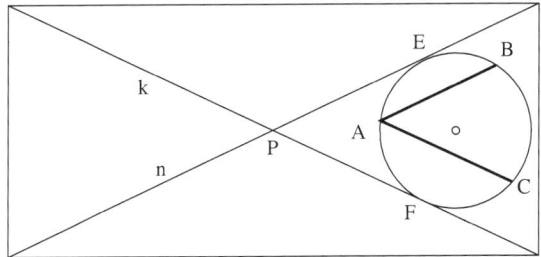

FIGURE 9.20

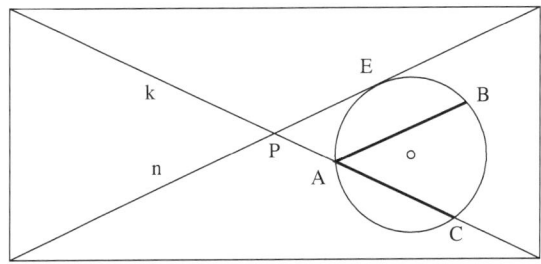

FIGURE 9.21

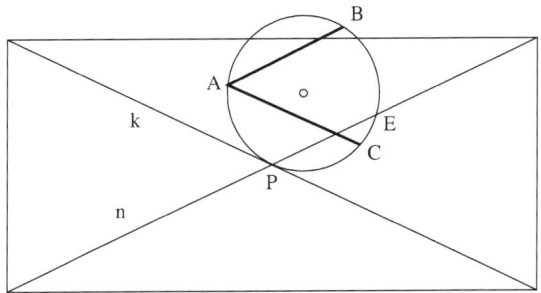

The crucial arc equality this time is $m\widehat{AP} = m\widehat{CP}$, and $m\widehat{AP} = m\widehat{BE}$. We begin as before:

$$m\angle P = m\angle A = \frac{1}{2}m\widehat{BEC} = \frac{1}{2}\left(m\widehat{BE} + m\widehat{EC} + m\widehat{PC} - m\widehat{AP}\right)$$

$$= \frac{1}{2}\left(m\widehat{EC} + m\widehat{PC}\right) = \frac{1}{2}m\widehat{PCE}.$$

This activity can also be done quite nicely with a computer drawing program such as Geometer's Sketchpad.

You should remember that although this was presented in the form of a demonstration lesson, it could very well be adopted as a student activity, where students working in small groups—each working with a cardboard model—do their own investigation of the measures of the various angle positions mentioned above.

Topic: Concept of Similar Triangles

Materials or Equipment Needed
A pantograph.

Implementation of the Motivation Strategy
The pantograph (figure 9.22) is a linkage instrument that is used to draw similar figures. It can be obtained in toy stores or drawing stores as it has been sold to children to allow them to draw

FIGURE 9.22

(Taken from www.rockler.com/product)

(or trace) cartoon characters in an enlarged form. It can also be constructed with strips of cardboard and fasteners.

The pantograph consists of four bars hinged at four points, with one point fixed and another point having a pencil. Holes are provided on the bars to allow for size adjustment. Tracing triangles and noting the ratio of similitude by the adjustments on the bars will be a good lead-in to the study of similarity.

Topic: Introducing Regular Polygons

Materials or Equipment Needed
Strips of paper or a roll of narrow (approximately 1 inch) paper strips to be distributed to the class, and some string.

Implementation of the Motivation Strategy
Regular polygons are familiar to most students, since they are the most prevalent figures in our environment. Yet, to introduce these through some clever paper folding might serve to provide a new level of appreciation for these geometric figures, one which should motivate much of the further study planned.

Begin by having the students tie the knots shown in figure 9.23 (a)–(d). For each knot, the students should leave the knot loose so that it can be analyzed and copied later.

Begin with the knot shown in figure 9.23 (a). This knot should now be copied with a strip of paper, pulled taut and pressed flat. The resulting regular pentagon should now be obvious. When held up to the light, the diagonals (forming a regular pentagram) are visible (see figure 9.24 (a)).

All regular polygons with an odd number of sides can be constructed by paper folding in this way. The knot shown in figure 9.23 (b) produces a regular heptagon (figure 9.24 (b)).

The regular polygons with an even number of sides are produced by two strips of paper following the model of the analog with two pieces of string. The construction for the regular hexagon is shown in figure 9.24 (c).

The construction of the regular octagon is a bit more difficult. The model knot is shown in figure 9.23 (d). Yet the paper folding

FIGURE 9.23

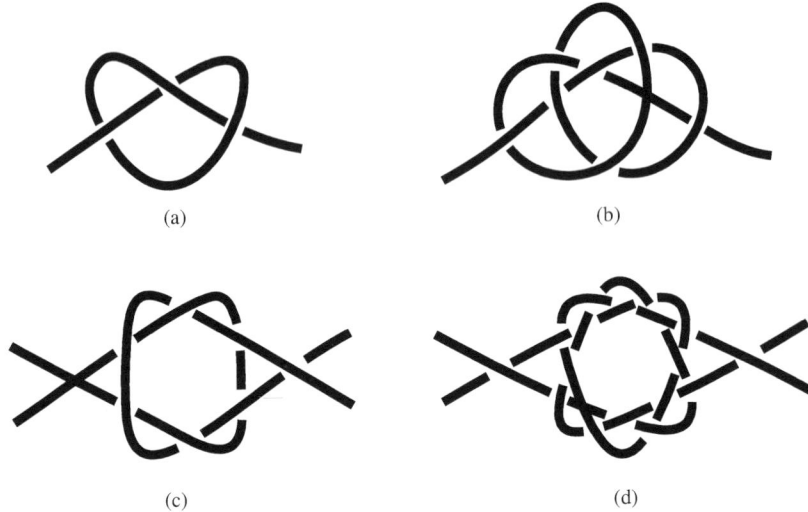

FIGURE 9.24

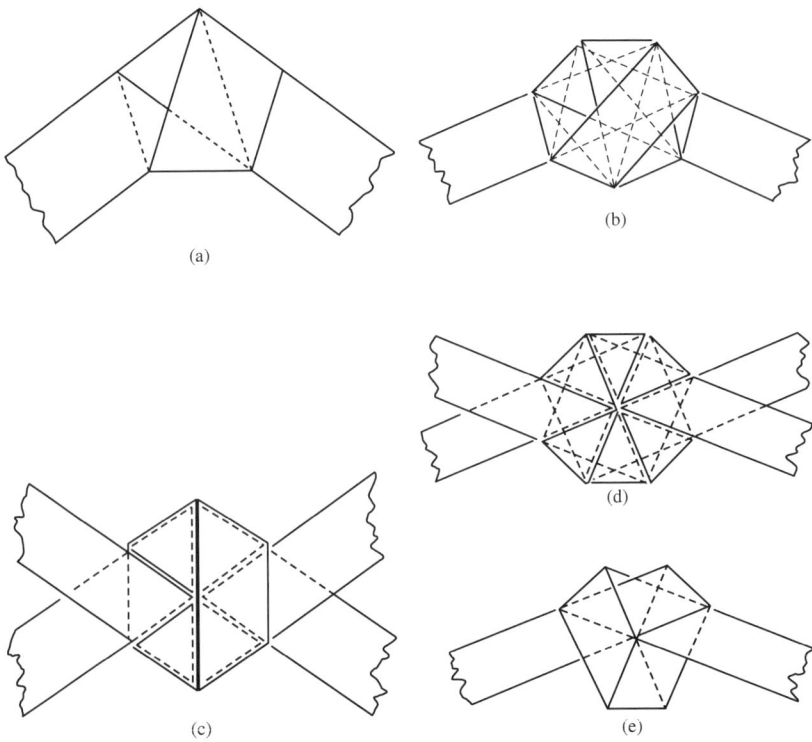

would be much easier if one strip were first folded as shown in figure 9.24 (e) and then interlinked with a second one of the same kind to form the octagon as shown in figure 9.24 (d).

This rather unusual opening to the study of regular polygons can broaden the experiential basis for students of geometry to the point where their thinking might produce some rather interesting results.

Topic: Introducing the Parabola

Materials or Equipment Needed
Waxed paper should be distributed to each student.

Implementation of the Motivation Strategy
The teacher entering the classroom with a stack of waxed paper sheets ready to be distributed to the class will generate an interest among the students—this in itself could be a motivating device!

A revealing approach to understanding the properties of a parabola is to have students generate, or construct, a parabola by means of the envelope of tangents to each curve. The basic idea used here is that a parabola is the locus of points equidistant from a fixed line (called the directrix) and a fixed point (called the focus) not on that line. We shall generate each tangent by applying this locus rule. This will be done by folding a sheet of paper and noting the pattern of the creases that result. (If waxed paper is not available, then any thin piece of paper may be used).

We show in figure 9.25 how to do it. Select a line segment, $\overline{AB}$, as the *directrix*. Choose a point, P, not on $\overline{AB}$, to serve as the *focus*. Fold the paper so that point P coincides with any point on $\overline{AB}$. Make a careful and distinct crease. You now have "constructed" the perpendicular bisector of a segment joining P and a point on $\overline{AB}$. Repeat this procedure by "placing" point P at a different position on $\overline{AB}$. Make another crease. Continue this process until P has been made to coincide with many different points on $\overline{AB}$. The more creases made, the clearer the parabola that will emerge from the tangent lines. It will not be necessary to actually draw the parabola, merely to observe how the many lines appear to be tangent to this (not drawn) parabola.

FIGURE 9.25

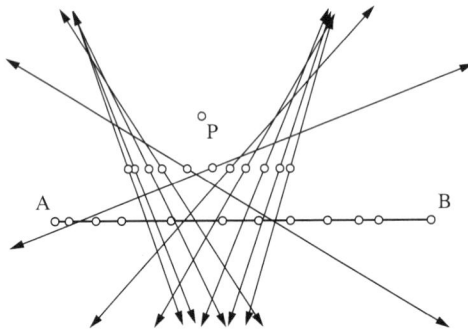

The drawing above, which indicates the result of folding the waxed paper, was done with the Geometer's Sketchpad computer program. By doing the construction on the computer, we have the ability to distort the diagram and inspect what aspects of the construction remain constant under various positions. Here we can see that the midpoints of the segments joining P with its various images on $\overline{AB}$ are collinear and that line is parallel to $\overline{AB}$.

The finished product makes an excellent bulletin board display, as well as a model suitable for discussing this conic section. Concepts such as the focus and directrix of a parabola are easily brought out in the process of the actual foldings of the envelopes of the tangents to the curves. A similar approach can be used to introduce other conic sections: the ellipse and the hyperbola.

Notes

1. Actually a gun and bullets would be a better analogue than the bow and arrow, since there the bullet can *really* be used only once. For this illustration, make it clear that the arrow, once shot, cannot be used again.
2. Obviously, in reality an infinite number of arrows would be required, so it must be appropriately simulated.
3. The book, *The Pythagorean Proposition*, by Elisha S. Loomis, contains 370 different proofs of the Pythagorean Theorem and was originally published in 1940, and republished by the National Council of Teachers of Mathematics in 1968.

10

Using Technology to Motivate Mathematical Relationships

With the advent of the ever-advancing technology there are many innovative uses to which this medium can motivate students. Perhaps the most compelling application would be to use dynamic geometry programs such as Geometer's Sketchpad, GeoGebra, or Cabri. Here, we can use the computer to have students construct some given configuration and then be given instructions to manipulate the construction and notice specific changes or constants. These should then lead students to question if these observations are always true or just in some instances that they may have witnessed with their particular activity. This leads nicely into a consideration of justifying observed phenomena. It should be noted that some of the ideas that are best suited with computer assistance were previously (and primitively) done with other devices. However, the computer makes these suggested motivational activities far more effective.

The computer can also be used to allow students to discover mathematical phenomena in areas such as probability, where, for example, the National Council of Teachers of Mathematics (NCTM) has a very broad website support called *Illuminations* (https://Illuminations.nctm.org), which provides many applications that can be used to motivate a class for a specific topic.

However, cautious selection must be exercised, since many of these units are intended to present a topic, while here we are only interested in using these units to motivate students towards the topic to be taught in the lesson following this brief activity.

Topic: Introducing Concurrency in Triangles

Materials or Equipment Needed
Geometer's Sketchpad, GeoGebra, or Cabri.

Implementation of the Motivational Strategy
Begin the lesson by having students draw a random triangle using one of the dynamic geometry drawing programs. Then have them locate the midpoints of the sides and draw the medians. Of course, they will see that the medians are concurrent. However, now have them distort the original triangle by moving one of the vertices with the cursor. They will notice that the three medians will always remain concurrent. Why is that? That will lead them to a desire to prove that this is, in fact, always true.

Topic: Introducing the Properties of the Midline of a Triangle[1]

Materials or Equipment Needed
Computer with a dynamic geometry program such as Geometer's Sketchpad.

Implementation of the Motivation Strategy
Have each student draw a quadrilateral—preferably not a square, rectangle, or any other parallelogram. Then have the students locate the midpoint of each of the four sides of their quadrilateral. They then should connect the midpoints sequentially. Have the students show their resulting construction to the class. They will be amazed that every student will have drawn a parallelogram!

FIGURE 10.1

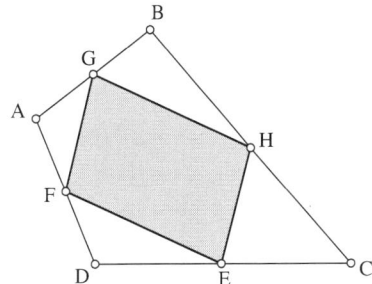

FIGURE 10.2

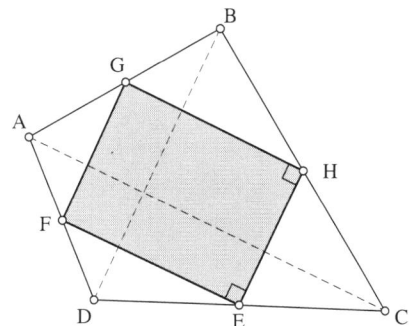

This resulting parallelogram (figure 10.1) is called a *Varignon parallelogram*.[2] This is truly amazing, since it will hold true for any shape of a quadrilateral, allowing all the students to have drawn a parallelogram. This should clearly arouse a curiosity among the students, thereby providing a fine motivation to justify why this is true. This will require learning about the midline of a triangle—the lesson of the day.

Once the students have proved that the midline of a triangle is half the length of the third side of the triangle and parallel to it, it is relatively easy to prove that the quadrilateral *EFGH* is a parallelogram. This is done by drawing $\overline{AC}$ and establishing that midline $\overline{GH}$ of triangle *ABC* is parallel to $\overline{AC}$ and half its length (see figure 10.1). The same can be done for $\overline{EF}$. Therefore, $AC = 2EF$ and $\overline{AC}$ is parallel to $\overline{EF}$, establishing parallelogram *EFGH*, since $\overline{GH}$ and $\overline{EF}$ are equal in length and parallel.

Of course, some quadrilaterals—when the midpoints of their sides are connected—will yield special parallelograms, such as a rectangle, a square, and a rhombus (a parallelogram with all sides equal). For example, suppose we consider a quadrilateral whose diagonals are perpendicular (see figure 10.2). When the side-midpoints of this quadrilateral are connected, a rectangle *EFGH* is produced.

If the quadrilateral has perpendicular diagonals that are also the same length, then the quadrilateral produced by joining the

FIGURE 10.3

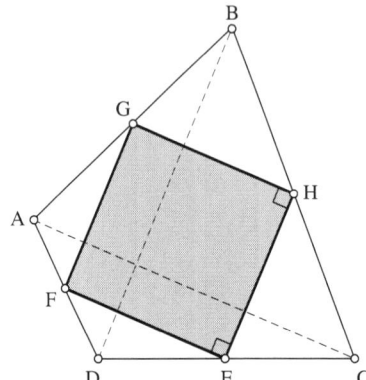

FIGURE 10.4

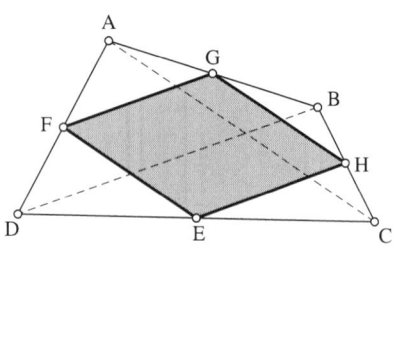

side-midpoints is a special kind of rectangle that we know as a square! (See figure 10.3.)

If the original quadrilateral only has diagonals of equal length and which are not perpendicular, then the quadrilateral formed by joining the side-midpoints will be a rhombus, as in figure 10.4.

In all, what began as a surprising curiosity and motivated the students to learn about the midline of a triangle, can lead to a nice investigation of quadrilateral properties.

Topic: Introducing a Proof Requiring a Transformation

Materials or Equipment Needed
Geometer's Sketchpad, GeoGebra, or Cabri.

Implementation of the Motivational Strategy
Using dynamic geometry software, students should draw any randomly constructed triangle and then construct an equilateral triangle on each of the three sides of the triangle. They should then join each vertex of the original triangle with the remote vertex of the equilateral triangle of the opposite side, as shown in figure 10.5.

FIGURE 10.5

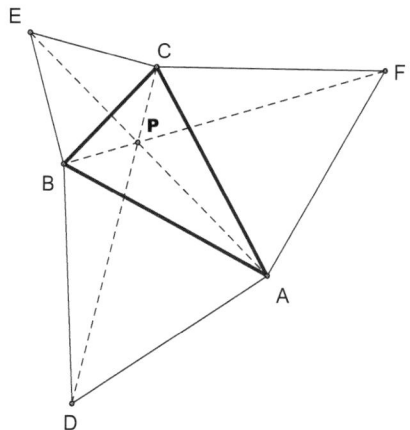

FIGURE 10.6

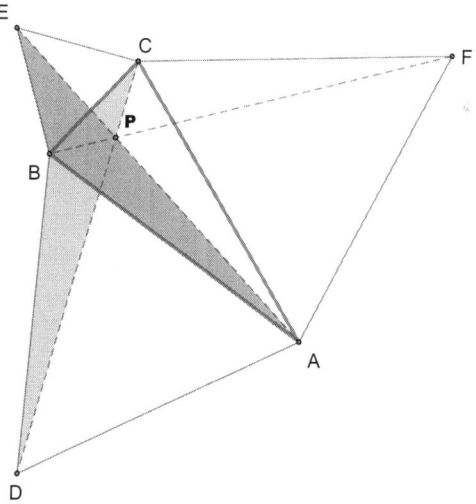

Students, when asked to change the shape of the original triangle will notice that the three dashed lines will always be congruent and concurrent. Why is that true? Will that always be the case? If so, how can we prove that in the general case? This will require proving $\triangle ABE \cong \triangle DBC$. (See figure 10.6.)

One approach to doing this is to show that rotating the triangles at point B will have them overlap and establish their congruence. This will be the lesson to be motivated.

Notes

1. The midline of a triangle is the line segment joining the midpoints of two sides of a triangle.
2. Named for Pierre Varignon (1654–1722), a French mathematician, who discovered it in 1713. Yet, it was first published (posthumously) in 1731. The Varignon parallelogram has half the area of the original quadrilateral, and its perimeter is equal to the sum of the lengths of the diagonals of the original quadrilateral. That is, in figure 10.1 the area of parallelogram *EFGH* is one-half of the area of quadrilateral *ABCD*, and the perimeter of parallelogram *EFGH* equals $AC + BD$.

11

Using Practical Problems to Motivate Instruction

It is not difficult to find everyday problems that can be used to motivate an ensuing lesson. However, one must be careful to be sure that the application selected is appropriate for the class audience. To give a problem about mowing a lawn to an inner-city class may be a waste of time, since they may not know what this is all about. Furthermore, selecting an everyday problem should be one to which the students can relate and which they would find useful to solve. Again, one must bear in mind that this problem should lead directly to the lesson planned.

Topic: Introducing the Pythagorean Theorem

Materials or Equipment Needed
A chalkboard or a whiteboard.

Implementation of the Motivational Strategy
Present the following problem to the class. You are in a furniture shop about to buy a circular tabletop, which measures 7½ feet in diameter. Your concern is whether this tabletop will fit into your house with a door frame that measures 3 feet wide and

7 feet high. Clearly, rolling this tabletop through the doorway vertically will not work, since the tabletop is a half a foot longer than the height of the doorway. The question is, if we tilt the tabletop, will it then fit through the doorway? We would then need to find the length of the diagonal of the rectangle that represents the doorway whose dimensions are 3′ × 7′. This will lead nicely into an introduction of the Pythagorean theorem.

Topic: Introduction to the Right Isosceles Triangle

Materials or Equipment Needed
A picture of a baseball field. The infield of the baseball diamond is clearly marked. The figure is a square with the four bases labeled first base, second base, third base, and home plate. The bases are each 90 feet apart.

Implementation of the Motivation Strategy
Tell the following story.

> A runner is standing on first base and decides to steal second. As he leaves first base, the catcher at home makes a fast throw from home plate to second base. The runner is out! How far was the catcher's throw?

Since the infield is in the form of a square, the throw from home plate to second base is the diagonal of the square. The figure formed by joining home plate to first base, first base to second base, and second base directly to home is an isosceles right triangle. The length of the hypotenuse is the distance we want to find. Have the class use the Pythagorean theorem to find the length, which turns out to be 127.28 feet. However, this is where students can be told that there is an easier way to get this answer, by knowing the special relationship of the sides of an isosceles right triangle.

This leads to a lesson that reveals the properties of a right isosceles triangle. The hypotenuse of a right isosceles triangle can be found by multiplying the length of the leg by $\sqrt{2}$. Or, given the length of the hypotenuse, the length of the leg is one-half the hypotenuse times the $\sqrt{2}$. The lesson would then progress to a consideration of the relationship between the sides of a 30-60-90 triangle.

Topic: Introduction to Mean, Median, Mode

Materials or Equipment Needed
A table of approximate average monthly rainfall for Naples, Florida as shown in figure 11.1.

Implementation of the Motivation Strategy
Ask the class whether or not they consider Naples, Florida, to be a rainy city. That is, what is the monthly *average* rainfall for Naples? This can lead into a class discussion of what is meant by an "average"—which is the topic of the ensuing lesson.

Actually, there is more than one "average" rainfall. An average is really a measure of central tendency. If we consider the month with most rainfall, we obtain July, which is approximately 9.2 inches. This would make Naples a very rainy city, indeed. If we examine the median, another type of average, which is the middle value, we obtain 3.0 inches of rainfall, which is mean between the two middle values—2.7 and 3.3—when the values are arranged in order of magnitude. The mode, yet another average, which is the value that appears the most number of times, is 2.0 inches of rainfall (January and November). The arithmetic mean, our most common average, is 4.65 inches of rainfall. Depending on which average we decide to use, the city of Naples in Florida could be relatively rainy. The following lesson will be an exploration of the three types of average.

FIGURE 11.1

January	February	March	April	May	June	July	August	September	October	November	December
2.0	2.3	2.2	2.7	3.3	8.9	9.2	9.0	8.7	3.8	2.0	1.7

Topic: Factors of Prime and Composite Numbers

Materials or Equipment Needed
A list of the counting numbers from 1 through 50, arranged as shown in figure 11.2.

Implementation of the Motivation Strategy
This motivational activity can be used to begin a lesson and class discussion of prime and composite numbers as well as the factors of numbers. Divide the class into two teams of equal numbers. Teams will alternate turns selecting a number from those left in the array. The number picked is added to that team's score. However, the opposing team gets a score equal to the sum of all the factors of the number picked.

For example suppose team #1 selects 50. Their score is thus 50 points. Now, team #2 gets all the factors of 50 left in the array, namely 1, 2, 5, 10, 25 for a total of 43 points. These numbers are then erased from the array. In this example it would be team #2's turn to select a number, and team #1 would get all the factors of that number. The array would now look like that shown in figure 11.3.

It's now team #2's turn to choose a number. Suppose they select 49. Their score is now 43 + 49. = 92. Team #1 gets the

FIGURE 11.2

1	2	3	4	5	6	7	8	9	10
11	12	13	14	15	16	17	18	19	20
21	22	23	24	25	26	27	28	29	30
31	32	33	34	35	36	37	38	39	40
41	42	43	44	45	46	47	48	49	50

FIGURE 11.3

		3	4		6	7	8	9	
11	12	13	14	15	16	17	18	19	20
21	22	23	24		26	27	28	29	30
31	32	33	34	35	36	37	38	39	40
41	42	43	44	45	46	47	48	49	

factors of 49 that are still in the array. The factors of 49 are 7 and 1. But 1 is gone, so team #1 gets only 7. This is a good pick. It lends itself to an investigation of the factors of perfect squares—always exactly 3). But would 47 be a better pick? Since 47 is a prime, the only factors are itself and one. So, team #1 would get 0 for the sum of its factors.

The game continues with the teams alternating picks and their opponents getting the factors of that pick. The number and its factors are erased from the array after their choice. The game ends when all numbers are gone and the array is empty.

Although this game may take more time than is desired for a motivational activity, there is also a built-in value to this approach in that students are becoming more familiar with number properties as they get into a lesson on prime and composite numbers.

Topic: Introduction to the Area of a Non-right Triangle

Materials or Equipment Needed
A diagram of a triangle with the side-lengths marked 30, 40, 70.

Implementation of the Motivation Strategy
Tell the following story.

> An advertisement appeared in a local newspaper making the following offer.
>
> ### FOR SALE!
>
> A triangular plot of land is offered for sale at the amazingly low price of $20 per square foot. The dimensions of the triangular piece of land are 30 feet, 40 feet, 70 feet. How much would we have to pay for the plot of land?

The students will try to figure out the square footage of the plot of land. Some may think of it as a right triangle and use the formula $Area = \frac{1}{2}(\text{leg})(\text{leg})$. However, there is no reason to assume the triangle is a right triangle. This is an excellent opportunity to introduce the formula for the area of a triangle given only the length of the three sides. This should lead into the lesson that would introduce using the formula for obtaining the area of a triangle, which is attributed to Hero of Alexandria (AD 10–70). It is as follows: $Area = \sqrt{s(s-a)(s-b)(s-c)}$, where $s = \frac{a+b+c}{2}$, known as the semi-perimeter.

Well, after the calculation, the area turns out to be 0.

Regardless of what the students try to do, it will take a few moments for them to realize that the triangle cannot exist. This leads into a lesson on the definition of a triangle, such that the sum of any two sides must be greater than the third side. The triangular piece of land cannot exist—the price would be $0.

Once that is done, before the lesson ends students may want to try to apply Hero's formula to a triangle that does exist. They might want to find the area of a triangle that has side lengths of 20, 21, and 70.

Topic: Introduction to the so-called Hinge Theorem

Materials or Equipment Needed
A chalkboard, or whiteboard is all that is needed.

Implementation of the Motivation Strategy
Tell the following story.

> Two friends are flying from the same airport but in opposite directions. Al flies 120 miles due west, then turns 45° in a southwesterly direction and then flies another 85 miles. Steve starts out due east, flies 85 miles then turns 90° towards the south and flies an additional 120 miles. Who is farther from the starting point?

The initial reaction of many students is that they are the same distance away, having both flown the same 205 miles from the airport. They should be encouraged to make a careful drawing of the situation to see what is actually happening. A carefully made drawing (roughly to scale) as in figure 11.4 will show that Al is farther from the airport than Steve.

FIGURE 11.4

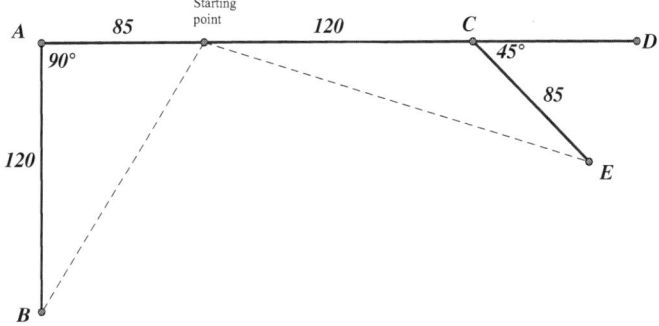

This leads into a lesson about the so-called "Hinge Theorem." (Some books refer to the triangle inequalities theorem as the "Hinge Theorem.") Basically, this motivator can be used to create a need for the following theorem: "If two sides of a triangle are congruent to two sides of a second triangle respectively, and the included angle of the first triangle is larger than the included angle of the second triangle, then the third side of the first triangle is larger than the corresponding side of the second triangle.

Topic: Introduction to the Odd-Even Properties of Numbers

Materials or Equipment Needed
A picture of a dart board as shown in figure 11.5.

Implementation of the Motivation Strategy
Tell the following story.

> Mary was playing a game of darts. She hit the target shown in figure 11.5 with exactly four darts. She said her score was exactly 61. How did she do it?

The students should begin to add different sets of four numbers each. They may get a score of 60 or 62, but not 61. It might take a few minutes. If no one realizes it, point out that all the numbers

FIGURE 11.5

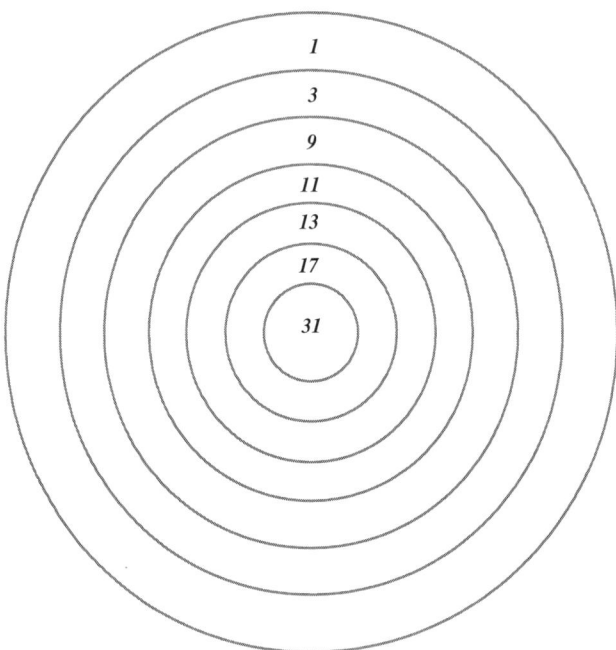

on the dart board are odd. It is impossible to obtain an odd score by adding four odd numbers.

This leads to a lesson that reveals the properties of operations on odd and even numbers. This practical application should motivate arithmetic operations with a concentration of number parity.

Topic: Introduction to the Sum of an Arithmetic Progression

Materials or Equipment Needed
A whiteboard or chalkboard.

Implementation of the Motivation Strategy
Tell the following story.

> Charles works in a supermarket. One day his boss gave him a box of 169 oranges and told him to stack them in a pyramid with 25 on the bottom "row," then 23 in the

next row above that one, then 21 in the next row and so on until the final and top row contained 1 orange. How many rows were in Charles' pyramid?

This leads to a lesson that deals with the sum of an arithmetic progression, $S = \left(\frac{n}{2}\right)(a+1)$, where n is the number of terms in the progression, a is the first term and l the last term. Thus, for Mike's problem, we need to find the value of n, when

$$a = 1, l = 25 \text{ and } S = 169$$

$$169 = \left(\frac{n}{2}\right)(1+25)$$

$$169 = 13n$$

$$13 = n$$

Notice that this is related to the Gauss story in Chapter 7.

Index

Note: italic page numbers indicate figures and tables; numbers in brackets preceded by *n* are chapter endnote numbers.

algebra: "coins in the dark" problem 93–4, *94*; and digit problems 97–9, *99*, 123–6, 127–8; and division by zero 59–60; exponential equations 22–4; "mind reading" activity 71–2; presenting challenge in 42–3, *43*; and proof of Thales' Theorem 57; and reducing terms of fractions 90–3, *92*
amazement/awe *see* "gee whiz" factor
amicable numbers 85–6
apple-gathering problem 53
applied mathematics *see* real-life situations; usefulness of topics
Archimedes 114–15
arithmetic mean 159
arithmetic series, sum of 107–8, 164–5
arrow analogy 132, 133, 150(*n*1)
autonomous learning 4
averages 159

base-2 number system 126–7, *126*

baseball field 158–9
Bible 104–5
birthday problem 6, 54, 120–2
bow-and-arrow analogy 132, 133, 150(*n*1)

Cabri 151, 152, 154
calculators xii, 21, 96, 103
Cantor, Georg 106
centre of gravity 110–12, *111*, *112*
challenges, presenting 6, 37–52; and algebra 42–3, *43*; benefits of small groups for 37–8; and circumference of circle 46–7, *47*; and concept of pi 44–6, *44*, *45*; and congruent triangles 48–50, *49*, *50*; and geometric series 51; and interior angles of polygons 47–8, *48*; and order of operations 38–40, *38*; and prime numbers 41–2, *41*; as rewarding experience 37
circles 15–18; and angle measurement 142–6, *143*, *144*, *145*; angles formed by secants inside 15–16,

15, 16, 143; angles formed by secants outside 16, 16, 143, 144–5, 144, 145; area of 65–6, 65, 133–5, 134; circumference of 44–7, 44, 45, 47, 77–8; diameter of, and cross-chords theorem 79–80, 80; external angles related to 16, 16; and Pascal triangle 32–3, 32; tangent segments to 17–18, 17; and Thales' Theorem 56–8, 56; *see also* pi
circumcenter of triangle 80–1
Common Core Standards 9
competency development 3
complex numbers 21–2; *see also* imaginary numbers
composite numbers 41, 42, 51(n2), 160–1, 160
computer technology xii, 7, 151–6; benefits of 151; caution in use of 152; and proof requiring transformation 154–5, 155; and triangles, concurrency in 152; and triangles, midline in 152–4, 153, 154, 156(n1); *see also* Geometer's Sketchpad
concentric circles 65–6, 65
concurrency, point of 81
conic sections 150
conjectures 105–6
core curriculum xi–xii
cosine function 13, 129–30
counter-intuitive conundrums 44–5, 53–4; algebraic 93–4;

apple-gathering problem 53; birthday problem 6, 54
counting combinations 34–6, 35
counting techniques 27–9; and palindromic numbers 28–9
cross-chords theorem 79–80, 80
curiosity 4, 6, 12, 19, 37, 53, 131; *see also* "gee whiz" factor; mathematical curiosities
curriculum, changes in xi–xii, 9
cylinders: and circumference of circle 44–6, 44, 45, 77–8; volume of 76–8; and volume/surface area of sphere 114–16, 115

da Vinci, Leonardo 138
dart board 163–4
data, organizing 28–9
de Polignac's conjecture 58–60, 59
definitions in mathematics 61
Desboves, A. 106
digit problems 122–6; in algebra 97–9, 99, 123–6, 127–8
directrix 149, 150
divisibility rules 95–7, 103, 122–3
division by 11 96–7, 99(n3)
division by zero 60–1
domain 133

earth, circumference of 46–7, 47
ego-related goals 3
Egypt, ancient 72, 141–2

Einstein, Albert 138
eleven: division by 96–7, 99(*n*3); multiplication by 95–6
ellipse 150
enthusiasm of teachers 1, 2, 4, 5
equidistance, calculating 81
equilateral triangle 49, 49–50, *49*, 135, 136, 154–5, *155*
Eratosthenes, Sieve of *41*, 42
Euclid 109, 138
expected outcomes 78–9
exponential equations 23–4

factors of numbers 85–6
factors, prime/composite 160–1, *160*
fair game 62
Fermat, Pierre de 50, 86
Fermat point 50
floor tiles 140
focus 149
formulae, deducing 26, 33; for area of circle 133–5, *134*; for area of triangle 18–19; for circumference of circle 44; for interior angles of polygon 48; for volume of cylinder 77
fractions 43, 89–93; rationalizing denominator of 89–90; reducing terms of 90–3, *92*
friendly numbers 85–6
function 132–3

Garfield, James A. 109, 138
gas meter 74, *74*

Gauss, Carl Friedrich 107–8, 165
"gee whiz" factor 6, 63–7; and apple-gathering problem 53; and area of circle 65–6, *65*; and area of parallelogram 63–5, *63*, *64*; and birthday problem 6, 54; and division by zero 60–1; and infinite series 66–7; and intrinsic motivation 54; and nature of proof (de Polignac's conjecture) 58–60, *59*; and nature of proof (optical illusions) 54–6, *55*; and probability, sample space in 61–3; and Thales' Theorem 56–8, *56*
geo strips 7, 131
GeoGebra 15, 33, 34, 74, 79, 80, 128, 129, 151, 152, 154
Geometer's Sketchpad 15, 33, 34, 63, 65, 74, 79, 128, 129, 130, 142–3, 146, 150, 151, 152, 154
geometric models 7, 131, 147–9, *148*
geometric series 51; infinite 66–7
geometry: altitudes of triangle, concurrency of 80–1, *81*; area of parallelogram 63–5, *63*, *64*; area of triangle 18–19; of circles *see* circles; concurrent angle bisectors of triangle 74–6, *75*, *76*; congruence of triangles 48–50, *49*, *50*; interior angles of polygons 33–4, *34*, 47–8, *48*; models for 7; visual nature of 2

Goldbach's Conjecture 51–2(n3), 105–6, *106*

Harpedonapts 141
heptagon, constructing model of 146, *147*
Hero of Alexandria 162
Heron's formula 18–19
hexagon: constructing model of 146, *147*; interior angles of 47–8, *48*
Hinge Theorem 162–3, *163*
hyperbola 150
hypotenuse 13–14, *13*, 158–9

imaginary numbers 20–2; and use of i 21–2
incenter of triangle 81
infinite geometric series 66–7
integers, non-positive exponents 29–30
intuitive thinking 17, 44, 46, 47, 62, 66, 79, 87, 122; *see also* counter-intuitive conundrums
isosceles triangle 57, 135, 136, 138, *138*; right 158–9

knots 147, *148*

large classes 8, 37
Leonardo da Vinci 138
lessons: class/work group sizes in 8, 37–8, 146; and enthusiasm of teachers xii–xiii, 1, 2, 4, 5; starting 4–5; teamwork in 160–1
Lincoln, Abraham 109

locus rule 149
logical thought 87–9, 119

mapping 133
maps 110–12, *111*, *112*
mathematical curiosities 7, 119–30; and base-2 number system 126–7, *126*; caution in use of 119; and digit problems 122–3, 127–8; and probability 120–2; and trigonometric angle sum function 128–30, *129*
mathematics: fear of 26, 84; history of *see* storytelling; lack of enthusiasm for 1, 2, 9
mean/median/mode 159, *159*
"mind-reading" activity 71–2
modular arithmetic 73–4, *74*
money calculations 70–1, 87, 103
motivation xii–xiii, 2–8; and amazement/awe *see* "gee whiz" factor; and autonomous learning 4; and competency development 3; and computers *see* computer technology; and discovering patterns *see* patterns, discovering; and entertainment *see* recreational mathematics; extrinsic/intrinsic 3–4, 54; five rules for 8; and justifying mathematical curiosities *see* mathematical curiosities, justifying; and novel events/activities 4;

and prepared materials *see* prepared materials; and presenting challenges *see* challenges, presenting; and real-life situations *see* real-life situations; and start of lessons 4–5; and stories *see* storytelling; students' lack of 1, 2; and task-/ego-/social-related goals 3; and usefulness of topics *see* usefulness of topics; and voids in students' knowledge *see* students' knowledge, voids in
multiplicand 95, 99(n1)
multiplication by 11 95–6

Napoleon Bonaparte 49, 52(n4)
Nicomachus 42
novel events/activities 4
number bases 123, 126–7, *126*
number sequences 30–3, *31*; Pascal triangle 32–3, *32*

octagon, constructing model of 147–9, *148*
odd-even properties of numbers 163–4, *164*
Oliveira e Silva, T. 106
one-to-one correspondence 133
optical illusions 54–6, *55*
order of operations 38–40, *38*
orthocenter of triangle 81

palindromic numbers 28–9; and logical reasoning 88–9

pantograph 146–7, *146*
parabola 149–50, *150*
parallelogram *14*, 63–5, *63*, *64*; and area of circle 134–5, *134*; varignon 152–3, *153*, 156(n2)
Pascal triangle 32–3, *32*
patterns, discovering 5, 25–36; in arithmetic series 108; and counting combinations 34–6, *35*; and counting techniques 27–9; and deducing formulas 26; and desire for structure 25; and interior angles of polygons 33–4, *34*; and non-mathematical problems 26–7; and non-positive integer exponents 29–30; and previous knowledge 25; stages of 26; and unusual/unexpected sequences 30–3, *31*, *32*
peer acceptance/approval 3
pentagon, constructing model of 147, *148*
percentages 69–70, 86–7
permutations 71, 72, 98, 99(n4)
pi 44–6, *44*, *45*; value of 103–5
Pipping, N. 106
place value 122–3
polygons: interior angles of 33–4, *34*, 47–8, *48*; regular, making models of 147–9, *148*
power 0, raising to 51(n1)
practical problems *see* real-life situations
praise 3
prepared materials 7, 131–50;

and angles related to circle 142–6, *143*, *144*, *145*; and area of circle 133–5, *134*; benefits of using 131–2; and concept of function 132–3; and parabola 149–50, *150*; and regular polygons 147–9, *148*; and similar triangles 146–7, *146*; and sum of angles of triangle 135–6, *135*, *136*; and triangle inequality 137–8, *137*, *138*
presidents of United States, birthdays of 120–2
prime numbers 41–2, *41*; and composite numbers 41, 42, 51(*n*2); and de Polignac's conjecture 58–60, *59*; definition of 41; factors of 41, 42, 160–1, *160*; function rule for 116–17; and Goldbach's Conjecture 51–2(*n*3), 105–6, *106*; and logical reasoning 88
probability: and birthday problem 6, 54, 120–2; empirical/experimental 79; expected outcomes in 78–9; fair game in 62–3; sample space in 61–3
proofs: of angles related to circle 142–6, *143*, *144*, *145*; and de Polignac's conjecture 58–60, *59*; and division by zero 60–1; justification of 7; and optical illusions 54–6, *55*; and probability 62–3; of Pythagorean Theorem 138, 150(*n*3); requiring transformation 154–5, *155*

proportions 70–1
props *see* prepared materials
punishment 3
puzzles *see* recreational mathematics
pyramids, height of (shadow reckoning) 72–3, *73*
Pythagoras 109, 138, 140, 141
Pythagorean Proposition, The (Loomis) 150(*n*3)
Pythagorean Theorem 13; extending 138–9, *139*; introduction to 108–10, *109*, 140–2, *140*, *141*; proofs of 138, 150(*n*3); in real-life situation 157–8

quadratic formula/equations 19–20, 22–3
quadrilaterals 14–15, *14*, 152, 153–4; centre of gravity of 111–12, *112*; *see also* parallelogram

rainfall 159, *159*
range 133
rational/irrational denominators 89–90
real-life situations 7–8, 157–65; and area of non-right triangle 161–2; and factors of prime/composite numbers 160–1, *160*; and Hinge Theorem 162–3, *163*; and interests of students 8; and mean/median/mode 159, *159*; and odd-even properties of numbers 163–4, *164*; and Pythagorean Theorem

157–8; and sum of arithmetic progression 164–5
recreational mathematics 6, 83–99, 119; and algebra 93–4, *94*, 97–9, *99*; benefits of 83–5; "coins in the dark" problem 93–4, *94*; digit problem 97–9, *99*; and divisibility rules 95–7, 99(*n*3); and factors of numbers 85–6; and fractions, rationalizing denominator of 89–90; and fractions, reducing terms of 90–3, *92*; level of difficulty in 84; and logical reasoning 87–9; and percentages 86–7; "think of a number" tricks 71–2, 84; *see also* mathematical curiosities
rewards 3
rhombus 154, *154*
Richstein, Jörg 106
right triangle 158–9; area of 18, 19; circumcenter of 81; sum of angles of 135, 136; and trigonometric angle sum function 128–30, *129*; *see also* Pythagorean Theorem

salary calculations 87
sample space 61–3
scalene triangle 49–50, *49*, 113–14, *113*, 136
shadow reckoning 72–3, *73*
Sieve of Eratosthenes *41*, 42
sine function 13, 129–30
sines, law of 112–14, *113*

small group settings 8, 37–8, 146
social-related goals 3
Sudoku 83
Solomon's Temple 104
spaghetti 137
sphere, volume/surface area of 114–16, *115*
sports examples/statistics 70, 158–9
square 14, *154*
square roots 21–3; of negative numbers 21–2; of quadratic equations 22–3
starting lessons 4–5
statistics 70
storytelling 6–7, 49, 101–17; and age of students 102; appropriateness of 102, 103; benefits of 101–2; and divisibility rules 103; and Hinge Theorem 162–3, *163*; and law of sines 112–14, *113*; and odd-even properties of numbers 163–4, *164*; and prime numbers 105–6, *106*; and prime numbers, function rule for 116–17; and Pythagorean Theorem 108–10, *109*, 140–2; and sum of arithmetic series 107–8, 164–5; timing/pacing in 102; and triangle, area of 161–2; and triangle, centroid of 110–12, *111*, *112*; and triangle, right isosceles 158–9; and value of pi 103–5; and volume/surface area of sphere 114–16, *115*

students: exploiting interests of 8, 69–70; exploiting voids in knowledge of *see* students' knowledge, voids in; families of, involving in lessons 70; lack of motivation of 1, 2, 9
students' knowledge, voids in 5, 11–24; and area of triangle 18–19; and circles, angle formed by secants in 15–16, *16*; and circles, external angles related to 16, *16*; and circles, tangent segments to 17–18, *17*; and desire for completion 11–12; and examples of familiar situations 12; and exponential equations 23–4; and imaginary numbers 20–2; and quadratic formula/equations 19–20, 22–3; and special quadrilaterals 14–15, *14*; and tangent ratio 12–14, *13*

tangent function 14
tangent ratio 12–14, *13*
tangent segments to circles 17–18, *17*
task-related goals 3
teachers' enthusiasm xii–xiii, 1, 2, 4, 5
team activities 160–1; *see also* small group settings
technology *see* calculators; computer technology
Thales and shadow reckoning 72–3, *73*

Thales' Theorem 56–8, *56*; proof of 57
"think of a number" tricks 71–2, 84
transparencies 7, 131
triangle inequalities theorem *see* Hinge Theorem
triangles: area of 18–19; area of, non-right 161–2; and area of parallelogram 63–5, *63*, *64*; centroid of 110–12, *111*, *112*; circumcenter/incenter/orthocenter of 80–1; concurrent altitudes of 80–1, *81*; concurrent angle bisectors of 74–6, *75*, *76*; concurrent medians in 152; congruence of 48–50, *49*, *50*, 154–5, *155*; and interior angles of polygons 34, *34*; midline of 152–4, *153*, *154*, 156(*n*1); Pascal 32–3, *32*; and shadow reckoning 72–3, *73*; similar 72–3, *73*, 146–7, *146*; sum of angles of 135–6, *135*, *136*, *see also* trigonometry
trigonometry: angle sum function 128–30, *129*; law of sines 112–14, *113*; tangent ratio 12–14, *13*

United States: center point of 110–11, *111*; presidents, birthdays of 120–2
usefulness of topics 6, 69–81; and algebra 71–2, 90–3, *92*, *93*–4, *94*; altitudes of triangle, concurrency of

80–1, *81*; and concurrent angle bisectors of triangle 74–6, *75*, *76*; and experiences of students 69–70; and modular arithmetic 73–4, *74*; money calculations/proportions 70–1; and probability 78–9; and rationalizing denominators of fractions 89–90; segments of circle 79–80; and similar triangles/shadow reckoning 72–3, *73*; volume of cylinder 76–8

varignon parallelogram 152–3, *153*, 156(*n*2)
videos 7, 131

Washington, George 109
worksheets 20, 38, *38*, 112, 113

zero: division by 60–1; raising to power of 51(*n*1)